The Pipeline and the Paradigm

SAMUEL AVERY

Ruka Press®

Washington, DC

First edition published 2013 by Ruka Press,® PO Box 1409, Washington, DC 20013.
www.rukapress.com

Library of Congress Control Number: 2013931105

ISBN 978-0-9855748-2-6

10 9 8 7 6 5 4 3 2 1

Printed in the United States of America

Design by Sensical Design & Communication

Table of Contents

Suncor tar sands mining operation, Alberta.

Foreword

By Bill McKibben

IT'S QUITE AMAZING HOW FAST THE KEYSTONE PIPELINE BECAME a symbol of environmental resistance.

In the late summer of 2011, as activists gathered in Washington, D.C., to begin civil disobedience outside the White House, the National Journal polled its "energy insiders" to see if they thought activism would slow the permitting process. Ninety-one percent of those Congressional aides, lobbyists, and journalists said no: the pipeline would have its permit by year's end.

Instead, President Barack Obama ended up denying TransCanada's request for approval—a temporary and partial win to be sure, but an indication nonetheless that people are learning to fight back against the fossil fuel industry.

In this case, that fight is being led by people on the ground—indigenous environmentalists in the tar sands country, and spirited resisters along the pipeline route all the way to the Gulf. And it's being chronicled mostly by people outside the mainstream media, able to use new forms to carry messages around the corporate power that dominates our traditional media. Sam Avery is the perfect example—from blogging to writing a book, he's gone deeper into the story of the pipeline than almost any other journalist. This volume tells a deep story about the resistance to the pipeline, one of the epic battles of recent environmental history.

The players are wonderful people, endlessly devoted: David Daniel, in Texas, say. Even as Obama denied permission for the northern section of the pipeline, he shamefully expedited approvals for the portions that crossed Texas—meaning the Daniel farm is still slated to be cut in half. So Daniel and his allies have been mobilizing for civil disobedience, joining hands with a wide swath of Texans (including members of the Tea Party) who are sick of corporate domination. Or heroes such as Nebraska's Jane Kleeb, who rallied the bright-red Cornhusker state against the pipeline, to the point where the state's

Republican senator and governor ended up opposing TransCanada, which offered Obama political cover as he made his call.

The Keystone fight is far from over. A second Obama administration may well give in to the fossil fuel industry. But there are encouraging signs elsewhere. In Canada, for instance, activists have rallied to block a similar pipeline west to the Pacific. And in Europe, environmentalists have fought, so far successfully, to make the EU classify tar sands oil as environmentally unsound.

It's easy to say: even if we stop this pipeline, we're still losing the climate change fight. And of course that's true—the temperature keeps rising. We really do have to take on the fossil fuel industry directly, try to figure out how to keep it from wrecking the climate. But have no doubt: the Keystone battle was the Lexington and Concord of this war. It sparked the largest civil disobedience action in thirty years on any subject; it produced the rarest of defeats for big oil. All we ever win in the environmental movement are temporary victories, and this one may be more temporary than most. But understanding its history is key to making sure there will be many more victories. Many thanks to Sam Avery for telling the on-the-ground truth, and for making sure we know what it all means.

Bill McKibben is an author, environmentalist, and activist. In 1988, he wrote The End of Nature, *the first book for a common audience about global warming. He is the co-founder and chairman of the board at 350.org, an international climate campaign that works in 188 countries around the world.*

The Pipeline and the Paradigm

Tar sands pumping facility, Alberta. (Jeff Whyte/Shutterstock)

THIS IS THE TIPPING POINT. WE ARE AT THE EDGE OF THE familiar world, standing at a moment of decision, staring into the abyss of runaway climate disaster.

There is enough carbon in the tar sands of western Canada to tip us over the brink. The Keystone XL pipeline will connect that carbon to the atmosphere. It will change the climate. It will diminish human habitation of the earth.

The pipeline is who we are now. It is our need. A deliberate step away from where we are—away from the brink—will open a universe larger than our need, with space to become a more gracious and viable form of life.

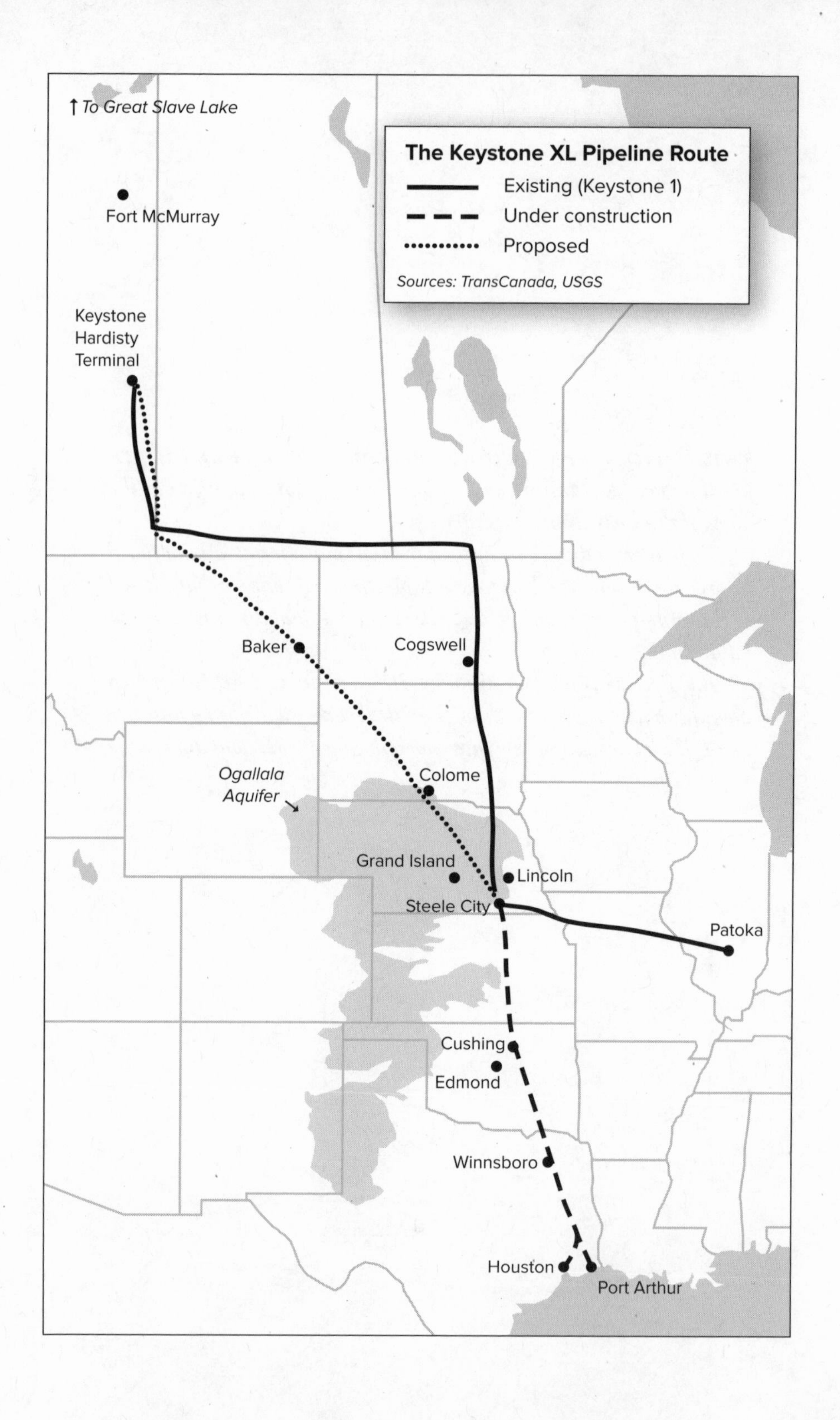

↑ To Great Slave Lake
The Keystone XL Pipeline Route
Existing (Keystone 1)
Under construction
Proposed
Sources: TransCanada, USGS
Fort McMurray
Keystone
Hardisty
Terminal
Baker
Cogswell
Ogallala
Aquifer
Colome
Grand Island
Lincoln
Steele City
Patoka
Cushing
Edmond
Winnsboro
Houston
Port Arthur

Introduction: Tar Sands and Mountaintops

February 14, 2012 – Frankfort, KY
High 38°F – Low 32°F – Precipitation 0.22 inches

Today the Senate votes. After two knockdowns from President Obama, the Keystone XL pipeline staggers back into the ring for another round. A presidential postponement late last year effectively canceled the project the first time. It came back as an amendment to the payroll tax bill, and Obama turned it down again. Now it has found its way back through the ropes to the floor of the U.S. Senate. In the last twenty-four hours, 802,120 people have signed a petition against it, including me. I have not yet heard the outcome of the vote.

IN FRANKFORT, KENTUCKY, THE DAY WAS FILLED WITH MARCHERS carrying signs and banners protesting mountaintop removal coal mining. Every February, a thousand people gather at the state capitol building to denounce the dynamiting of mountain ridges and the bulldozing of dirt and broken bedrock into valleys and streambeds for the sake of cheap fossil fuel. I was exhilarated, as always, by the lofty cheer of the crowd and frustrated, as always, by the unwavering apathy of everyone else.

But there was something very different this year. The keynote speaker was not from Kentucky. She had not been to Appalachia, and she did not know the particulars of mountaintop removal. Melina Laboucan-Massimo is a member of the First Nation Cree, an indigenous people whose land in Alberta, Canada, is being sacrificed for tar sands extraction. Bitumen mined from the tar sand by the billions of barrels will be piped along the Keystone XL pipeline from Alberta through Saskatchewan, Montana, South Dakota, Nebraska, Kansas, and Oklahoma to a deepwater port in Texas to be refined and sold on the world market. Melina spoke not of coal mines and mountaintops but of tar sands and pipelines. We knew exactly what she was talking about, though we knew little about the land she comes from. In their effects on the land, water, air, and local populations, tar sands extraction and mountaintop removal are one and the same. The deforestation is the same; the pollution is the same; the ill health, the distant profits, and the uprooting of rural communities are all the same. The people of the Cree nation and the mountain people of eastern Kentucky are separated by culture, geography, and an international boundary, but their everyday lives are encircled by the same struggle: fossil fuel extraction is literally undermining the ground they stand on. Some make their living from the process, as others watch their world slide away.

I am passionately opposed to mountaintop removal. The Appalachian hills of eastern Kentucky, West Virginia, Tennessee, and Virginia are home to the most diverse deciduous forest in North America. Poplar, oak, hickory, maple, beech, sassafras, and dogwood root into the dark mountain soil, their thick hardwood branches arching in the canopy overhead. Minerals flow from deep in the earth through root, trunk, and branch to leaves suspended high in the air, where sunlight turns them into new life. As leaves fall to the earth, life from the sun returns to the soil, sprouting into the shrubs and wildflowers of the forest floor. Mountain laurel, rhododendron, wild ginger, lady slipper, saxifrage, poison ivy, and Virginia creeper weave among the tree trunks, sipping dappled sunlight that slips between leaves overhead. Deer, rabbits, squirrels, bobcats, field mice, and raccoons sniff the air, searching for their next meal. Goldfinches, turkeys, tanagers, and cerulean warblers fill the branches, while dragonflies, butterflies, and cicadas zigzag across the open air. Ants, deer ticks, beetles, and centipedes creep through unseen crevices. The pileated woodpecker carves his home from a rotting hardwood trunk. The web of life is tightly woven in Appalachia, but to those who mine coal by removing mountaintops, the web of life is known as *overburden*.

Overburden is anything above the coal seam—anything that stands in the way of extracting coal from the mountains and bringing it to market. The wildlife is pushed back, the trees and flowers are bulldozed, and, day after day, tons of dynamite are blasted through the bedrock to break the overburden into pieces small enough to push over the mountainside into the streams below. Hundreds of miles of streambeds are buried, killing fish, frogs, and crustaceans. Rumbling shockwaves crack foundations of nearby homes, coal and rock dust cover cars and houses, and wells are ruined as water runs orange through the tap. The breaking up of so much bedrock vastly increases the surface area of rock subject to leaching, thereby releasing selenium, arsenic, copper, sulfur, sediment, electrolytes, and iron into the groundwater. Even natural gas gets into tap water—I have seen people turn on their faucet and light it with a match. The everyday lives of people who live in Appalachia have become part of the overburden.

After about a year of mining, the coal is gone. The massive shovels and draglines are gone; the coal trucks and bulldozers are gone; the forest and the jobs are gone. The community is left with a bare spot where a mountain used to be. There was a saying that coal is Kentucky's "ace in the hole." Now, all we have is the hole. If mountaintop removal were to happen near major cities or along interstate highways, the public would be so outraged it would end immediately. If a foreign country were to do this, we would consider it an act of war. But it happens in the hills of Appalachia, where most people do not see it. Even in Appalachia itself, mountaintop removal remains largely invisible, because you can't see it from your car. The roads and houses and stores are in the valley bottoms. You have to get out and walk over a ridge or two to see it. Or, you can see it from space. Check out Google Earth at 37 32' 30" N, 82 10' 40" W, or just Google-fly over eastern Kentucky and West Virginia and zoom in until you start to see white patches that look like lesions in the forest. Move around a bit, and you will be able to find any number of coal mines that used to be mountains. Look at the remaining forest between sites. You are not looking at anybody's description of mountaintop removal; you are not looking at a framed photograph or a documentary presentation. You are looking at the thing itself—at how the earth is beginning to appear from space.

Our speaker's passion at the rally was for the despoiled lands of her people in western Canada, thousands of miles away: a different country, a different culture, a different topography, and a different kind of fossil fuel. Yet she spoke to us, to our passion. It is the same.

Here is Fort McMurray, Alberta, the heart of the Canadian tar sands region: 56 57' N, 111 23' W. Take a look at the mines and tailings ponds. It's a moonscape.

We in Kentucky share with the Cree nation a continent and a planet. It is here that our interests converge.

The end product of mountaintop removal is, of course, coal. The useful portion of coal is almost entirely carbon. When burned, coal produces carbon dioxide plus energy. The energy is used mostly to generate electricity, while the carbon dioxide is released into the atmosphere. There it accumulates and traps solar energy, causing climate disruption. (Selenium, arsenic, and mercury go up in smoke as the coal burns and settle on the ground and in waterways.) A top NASA climatologist, James Hansen, estimates that a healthy level of carbon dioxide in the atmosphere should be no more than 350 parts per million. In 2012, the level is 393, and it's increasing by about 2 ppm every year. While

Tar sands mine near Fort McMurray, Alberta.

mining coal destroys the land and the water, burning coal destroys the air and the climate.

Tar sand bitumen is 5 percent sulfur, half a percent nitrogen, one-tenth of 1 percent heavy metals, and the rest hydrocarbon. The industry likes to call it "oil sand," but it is as sticky as tar on a cold day. It's sludgy rather than oily because, as a low-grade hydrocarbon, it has far more carbon than hydrogen, and it must be *upgraded* by adding hydrogen. The upgrading requires burning other fossil fuels, usually natural gas, to convert bitumen into synthetic crude oil, something that will move through a pipeline.[1] Upgrading is so energy intensive that it adds one to two hundred pounds of CO_2 to the atmosphere for every barrel of bitumen upgraded. Compared to refining conventional oil, upgrading tar sands (removing impurities and adding hydrogen) produces two to three times more sulfur dioxide (which causes acid rain), volatile organic compounds (producers of ozone), and particulate matter (a cause of heart and lung diseases).

Upgrading is where the carbon bomb begins to explode: a single bitumen upgrading facility using natural gas blows 1.3 megatons of CO_2 into the air every year. If bitumen itself is used as an energy source, twice as much, or 2.6 megatons, explodes into the atmosphere. A coal-fired tar sands upgrading plant is even worse—3.8 megatons a year, the same as eight hundred thousand private automobiles.

The Alberta carbon bomb arsenal blasted a total of 40 megatons in 2007. By 2020 it is expected to release 127 to 140 megatons a year.[2] And that's just upgrading bitumen to crude oil status. When refined and burned for fuel, the Canadian tar sands will add around 50 to 60 ppm of CO_2 to the atmosphere over the next 50 years, bringing the total atmospheric carbon levels to an Earth-toasting 450 ppm, without additions from other sources such as coal, natural gas, and conventional petroleum. This is in addition to the carbon released from upgrading and the loss of carbon sequestration in the lands where bitumen is extracted. The peat bogs and hardwood forests that are destroyed where tar sands are mined remove twice as much carbon from the atmosphere as tropical forestland.

The important number to remember here is 450 parts per million. This is the low-end estimate of carbon dioxide that will be in the atmosphere as a result of tar sands extraction and combustion over the next fifty years. The high-end estimate is 540 ppm. This includes the carbon costs of extraction, upgrading, and refining tar sand. (The pre-industrial level was 280; the "safe" level is 350; the current level is 393.) The 450 ppm number is important because it is also the likely threshold level at which critical climate feedback loops take effect and runaway global warming begins.[3]

ABOUT 20 PERCENT OF TAR SAND DEPOSITS ARE SHALLOW enough to be strip-mined, destroying the entire surface of the land. The other 80 percent is extracted *in situ*, a process that preserves most of the surface, but slices it up with pipelines, roadways, and seismic lines. The Cumulative Environmental Management Association, an industry-funded group, estimated that *in situ* bitumen mining would make one to three million acres of land unsuitable habitat for fish, bear, moose, and caribou. *In situ* extraction is also more energy intensive than surface mining, because the tar must be heated to be pumped out of the ground.

More than a hundred bitumen-extracting projects have been approved in Alberta already, making the overall operation the largest energy project in the world. The tar sands area contains a third of the world's known petroleum resources. It encompasses an area of 54,000 square miles, or about one-fourth the total area of Alberta. The mining process loads raw tar sand into 400-ton dump trucks, the largest trucks in the world, which dump the ore into a crusher that pours it onto the world's largest conveyor belt, nearly a mile long. Today, even before the XL pipeline, Canada produces more oil than Kuwait.

Tar sands extraction also uses—and pollutes—enormous quantities of water. It drains watersheds and destroys thousands of acres of wetlands by withdrawing groundwater at rates measured in the billions of cubic feet and trillions of gallons. Net water runoff in the Athabasca River has declined by 30 percent since 1971, and it is projected to decline to 50 percent by 2050. Twelve barrels of water are required to make one barrel of bitumen. This produces 400 million gallons *a day* of toxic wastewater at the tar sand mines. Leaky tailings ponds storing this wastewater already cover more than fifty square miles of former forestlands. Migrating waterfowl, including the whooping crane, cannot distinguish tailings ponds from natural lakes and routinely die in the poisoned water.

The greatest obstacle the bitumen industry has faced is moving the product to market. Northern Alberta is a long way from a major population center or a deepwater port. There are already pipelines moving bitumen south to the United States, but they are not as big as the Keystone (which will be 36 inches in diameter) and do not reach the Gulf of Mexico. The new pipeline is extremely important to the tar sands industry, because once the connection is made between the deposits in Canada and a deepwater port in Texas, the fuel will be available everywhere in the world. Promoters suggest that the oil will be for domestic American consumption, which is true, but it will also be sold in Japan, India, Europe, Africa, and China. Petroleum is bought and sold on the world market; once it is loaded in a supertanker, it can go anywhere. There are no earmarks on a barrel of oil. If it becomes available everywhere, it will be burned everywhere, and when it is burned everywhere, it will be, as Hansen has said, "game over" for the climate on planet Earth. It is enough to evoke passion in anyone: Cree, Appalachian, black, white, yellow, right-handed, left-handed, conservative, liberal, man, woman, or child. It will change everyone's life, and we are beginning to wake up to it. The movement to stop the pipeline will become part of the larger global movement by people everywhere to save the land, the water, wildlife, the climate, and themselves.

BUT WHERE PASSION BEGINS, REASON ENDS. THERE ARE GOOD reasons to remove mountaintops and to build pipelines. If passion is to become creative, it must be guided by reason, and it is important to know what the reasons are on both sides.

A year or two ago, I went with a group of friends to a hearing on coal mining permits in eastern Kentucky. We drove in several cars and parked close to one another, then pulled off our bumper stickers (*Save the Mountains, Clean Coal Isn't, Clean Coal is Like Dry Water, I Have Been to the Mountaintop—But It Wasn't There,* etc.). You need to be careful how you come across in this part of the world. The setting was a large basketball arena that seated five thousand people. Four thousand nine hundred and seventy of them were coal miners, equipment dealers, local businessmen, and politicians. The other thirty were us. We told ourselves, as we made our way through the crowd, that the miners were there because they had been given a paid day off and bused in at company expense.

But we quickly realized that they wanted to be there. Speaker after speaker got up and told how they had been miners all their lives, like their fathers and grandfathers, and proud of it. Storeowners told of customers with coal company paychecks, teachers told of school projects paid for with coal money, and city councilmen told of revenues streaming in from coal severance taxes. A candidate for the U.S. Senate stood up and claimed that the issue was not mountaintop removal, but mountaintop *development*.

We got up, too, and had our say, and they listened. But we were fighting for an intellectual, environmental nicety while they were fighting for jobs, livelihood, and dignity. They shouted, they cheered, they chanted; *they* were passionate. We heard about shopping centers, Walmarts, prisons, and country clubs built on newly leveled land. Herds of elk were gently grazing reclaimed mines, and big-dollar tourists were flocking in to see them. We heard that all the energy we would need was right here in America. Coal can be made clean: all you have to do is install scrubbers and sequester the carbon in old salt mines. Even if supplies of coal eventually run out, we need energy *now*, and coal is the best and cheapest way to get it. If you don't like coal, we were told, just pull the plug, sit in the dark, and see how you don't like it. Coal is our past, our present, and our future. One miner asked me, politely, what I did for a living, as if he were wondering if I did anything at all.

Jobs and energy security are the reasons for the tar sands pipeline, too. For the construction phase, supporters project around five thousand new employees in the United States, which, multiplied by demand for hotel rooms, meals, and entertainment, they say will lead to twenty or even fifty thousand jobs, all created by *private* investment. No tax dollars needed. Canada has the resource, right over the border. Should we get the petroleum we need from them or from the Middle East? Could we avoid another Iraq war? If we don't buy it from our Canadian neighbors, the Chinese will; another pipeline over the Rocky

Mountains will get it to the Pacific coast for direct shipment to Asia. All the carbon will end up in the air anyway; we might as well be the ones to burn it.

And there's another passion. Private money will lay the pipe, extract the tar, refine it into gasoline, and distribute it to those who want it. Market forces will attract capital to build the infrastructure, and then—best of all—force down the price at the pump. No bureaucracy needed. Private investment will keep government off the backs of the people. Citizens will have lower taxes and cheaper gas. But this is not just a matter of money; people are sick and tired of the government telling them what to do and how to do it. They want good, steady jobs making things people need. The pipeline would do just that.

THE KEYSTONE XL PIPELINE WILL BECOME A MONUMENTAL clash of economic and ecological interests. How you understand the tar sands pipeline depends on where you are looking *from* as much as what you are looking *at*. I realized, as I stood amid the signs and banners on the steps of the state capitol listening to Melina's horror story of bitumen mining in Alberta, that the pipeline will be a fight for who we think we are. The physical scale of the project, the reptilian appetite for fossil fuel, the gigadollars behind it, the jobs, the landowners, the aquifers, the spills, the forests, the birds, the cancer, the birth defects, and the destruction of the planetary climate are enough to make this the environmental issue of our time. The stakes are that high.

All the little pictures add up to a very big picture, a huge picture—the picture of how we will live or not live through the twenty-first century. But the little pictures do not add up by themselves; they do not merge automatically into a big picture. The backhoe driver with a family, the welder with a mortgage, the Cree fisherman with cancer, the car owner, the environmentalist, the banker with a billion-dollar investment, the rancher staring at an oil spill, the whooping crane circling to land in a polluted tailings pond, the mother and father thinking about the future—all of these pictures will collide in a tangled mass if they are not arranged into meaningful order. No single arrangement can be objectively better or truer than another, but it dawned on me, as I stood watching the crowd, that I might be able to make a unique arrangement. As an environmental activist, historian, carbon consumer, businessman, family man, solar installer, and all-purpose philosopher, I might be able to put the pieces together in a new way. I have seen some things. The picture would not be unbiased—I am too much the tree hugger to pretend editorial balance. But it would be a big picture, a picture extended into the past and future, deep into the soul of the living planet, and high into the firmament of earthly spirit. It would be a vision of human purpose transcending economic gain.

The picture stirred within my body. The spirit rose through my legs and chest and filled my mind as our speaker finished her message and a roll of applause thundered from the crowd.

I began to move. Thoughts churned through my head as I snaked through the signs and onlookers to the podium. I will have to shape this, sculpt it. It will not come out in words as I feel it now. I will go to Alberta. I will talk to Melina, see the picture she sees, lay my own eyes on the tar sand biocide of her native land and people. I will follow the pipeline and see the pictures in Nebraska and Oklahoma. I will know what it is like for the farmer in Texas and the rancher in South Dakota. I will seek out the activist, the landowner, and the truck driver. My feet will be on the ground, but my spirit will soar. I will see the earth as she sees herself.

What I want to do with this book is look into the minds and spirits of people, like me, who understand that the earth is in crisis and feel they have to do something about it. What is the spark within? What impels ordinary people to go forth, carry signs, attend rallies and hearings, write letters, walk miles, disturb the peace, and submit to arrest? What sort of mad citizen's disease grabs hold of people and moves them against the wisdom of their leaders? Why do otherwise sane people spend free time opposing directions society is taking for their benefit? And, for those building the pipeline, where does *their* passion come from? How could they possibly oppose our efforts to save the earth? Why is the divide between us so wide and so deep? How can reasonable people see such different worlds? Finally, I want to look at what is truly practical—at what actually can be done, and must be done, to maintain a human presence on Earth—a healthy, *employed* human presence.

February 14, 2012 – Frankfort, KY
High 38°F – Low 32°F – Precipitation 0.22 inches

I just heard on the radio that the Senate has decided to delay its vote on the pipeline. I also heard back from my senator, Rand Paul, regarding the petition I signed. Referring to the president's opposition to the pipeline, he said, "Rather than create thousands of new jobs, expand America's refining capacity, and strengthen our unique partnership with neighboring Canada, the President has elected instead to use this project as an opportunity to appease environmental extremists."

The president has vowed to veto any transportation bill that includes the Keystone XL pipeline. The Senate vote is likely to come up toward the end of February.

PART

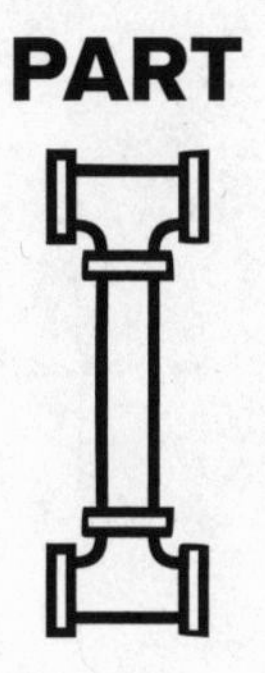

The Paradigm

The author, under arrest in Washington, DC, August 22, 2011. (Shadia Fayne Wood)

The Dance

LESS THAN A YEAR AGO, I RECEIVED THE FIRST E-MAIL ABOUT the pipeline. I had never heard of the Keystone XL. There was something up there in Canada somewhere called tar sand, I remembered, but it was just another one of those things, like whale hunting, dead coral reefs, or civil wars in sub-Saharan Africa, that line the back corners of my mind. I had not given it much thought. I concentrate my thinking and effort on things that I can do something about. I don't take the whole world on at once. Tar sands, I thought, were something like oil shale: a low-grade oil substitute far too expensive to refine into anything useful. And the deposits were probably so small as to be unworthy of more than passing attention.

Then I heard about the carbon bomb. I wasn't aware how enormous the tar sands were. I have been working for years on climate issues, concentrating on coal and mountaintop removal, and here was an entirely new phenomenon: the second largest carbon sink on the planet! I had no idea. There is enough of this stuff to boil the oceans!

Coal mining and oil drilling are global issues, too, but this particular issue has a better chance of being *perceived* as global because it crosses an international border. This is not Department of Interior material; this is State Department. The pipeline will be a presidential issue and get plenty of media attention. It will be seen. Unlike earth-gouging coal mines hidden in the back hollers of Appalachia or supertankers streaming unseen across oceans, the pipeline is going to stretch across the midsection of America, crossing farms, ranches, and interstate highways. People are going to know about this.

So I signed up. More than a dozen of us from 350 Louisville—our local, informal environmental organization—signed up to engage in civil disobedience in front of the White House. None of us had been arrested for civil disobedience before, though we did try once. Back in March of 2009, the same bunch of us, more or less, took part in national climate-change group 350.org's attempt to shut down the coal-burning Capitol Power Plant in Washington. Thousands surrounded the plant, blocked all the entrances, and shut it down, illegally. But the police just stood by and waited it out. They knew we wanted to be martyrs, and they weren't playing along. There were hundreds of them, too. They were polite to us, and we to them. We shouted and chanted and sang songs and listened to speeches, and they stood there in the cold until we left. I shouted, "Thanks for being here," to a group of them across the street and one shouted back, "Thanks for the overtime pay!" We had all been trained in what to bring, how to act, how not to act, and what to expect when arrested, but the police didn't cooperate. They wouldn't arrest us. So we went home. It's embarrassing to be ignored when you're trying to break the law.

I came close to being arrested at a 1970 demonstration in D.C. against the Vietnam War. We had been trained in nonviolent civil disobedience and were attempting to block traffic on bridges across the Potomac River. An officer looked me in the eye and shouted at me to stop, so I ran like hell. He wasn't at all polite. I was in a lot better shape than he, so I didn't have to be polite either. He chased me for a block or two and gave up. The only time I did manage to get arrested was two years earlier, in Ohio, for hitchhiking.

So the Tar Sands Action was going to be a new experience for me and for all of us from Louisville. This time they would have to arrest us: we would be blocking the postcard view of the White House near Lafayette Park. The police had to let us on the sidewalk, because they had to let the tourists there to snap photos. But they could not let us stay there. Because of large crowds and security concerns, tourists have only a minute or two to take their photos and then must move on. So when we sit down and refuse to leave, they will have to arrest us. They'll have to make us heroes this time.

The Tar Sands Action was planned by 350.org to be a series of daily actions at the White House. Wave after wave of protesters would submit to arrest over a three-week period. We signed up for day three: Monday, August 22, 2011. We drove from Kentucky on Sunday, arriving in time for civil disobedience training that night. Along the way, we heard on the radio that those arrested on Saturday were *not* released, as expected, and were still in jail. Nobody knew when they would be released. Was this to discourage the next wave, namely, us? We had not planned on being held—we had things to do back home. When you engage in civil disobedience, you want to know beforehand how long you will be held—that's natural—but the first thing you learn when you do it is to be ready for anything. Organizers and planners may give you a general idea of what to expect, but there are no guarantees. When you get arrested, you're no longer calling the shots. As soon as those handcuffs snap around your wrists, you're in for the duration, whatever that may be.

This is why being arrested for what you believe in becomes a transformational experience. It is not something you can plan your day around, not something that fits into the rest of your schedule. Everything else in your life comes to a stop. You step out of the normal flow of time—away from your house, your car, your family, your morning routines, and your workday—into an unknowable realm of waiting and wondering. You have no control. Someone else says where you go and what you do. The world no longer looks the same to you, and you no longer look the same to the world. You have been removed from society. After you reenter, you are never fully the same.

We arrived at Lafayette Park midmorning, listened to a few speeches, and lined up. There were fifty-two of us that day from all over the United States and Canada. The Kentucky contingent filed onto the sidewalk and sat down in front, holding a long banner. As we took our assigned positions, a swarm of photographers rushed in from the sidelines, knelt on the pavement in front of us, and began snapping pictures by the hundreds. The Tar Sands Action story had caught hold over the weekend, and the media was out in force. Our picture with the banner would appear in the *New York Times* and on the cover of the *Huffington Post.* It was a moment of heightened awareness: I can close my eyes now and feel the concrete beneath me, hear the shutters zipping, and see the cameras, the police, the paddy wagons, and the tourists behind them. Time slowed. I was glad to be me, doing what I was doing, right there, at that moment. My life had meaning.

Why do we have to get ourselves arrested? We're upstanding, educated, law-abiding, patriotic, family people with kids and jobs and productive things we could be doing. What does deliberately breaking the law have to do with pipelines and melting icecaps? Are we just trying to get attention? Well, yeah ... but we could be holding signs, or signing petitions, or writing our congressmen. Why do we have to cross the line to the *other side* to make a point?

The police moved in, and the cameras moved back. The morning was sunny and cool but warming fast. A young U.S. Park Police officer with a grim look brought a bullhorn to his face and issued the first warning: everyone must leave the area. Those who stay will be subject to arrest. The crowd fell silent as a handful of spectators and support people quietly tiptoed off the scene.

Why arrested? What were we doing here getting arrested? Why were we surrounded by people in uniforms threatening to put us in handcuffs? The police were here in full force to carry out the laws of society, to do what they are trained and paid to do. I help pay them, and I wanted them to do what they are paid to do. They were calm, professional, and grim. They had to be vigilant, ready for anything we may do. They had guns, clubs, tasers, and handcuffs. They could hurt us, even kill us. We let them do to us whatever they would do. We trusted them with our bodies. Sharpshooters with high-powered rifles looked down at us from the roof of the White House. Police in plainclothes milled through the crowd, blending in with the tourists and homeless people.

I sat cross-legged on the concrete, feeling the force of life stirring within the ground and moving up to the surface from deep within my soul. Breath and heartbeat rose from the earth. I did not imagine it; I felt it. This was me, now: life coming through this body and radiating out to the street. Eyes everywhere were on us; we had nothing to do but sit and wait, watching, looking. I was happy and calm. People near me whispered softly. The dance was about to begin. I would rise, learn the step, watch the choreography unfold, watch myself watching. But why did I need to be arrested?

My mind zoomed out from the street, the park, the pavement. My senses were heightened, but I was no longer seeing from my body. I was looking from high overhead at myself and fifty-one others in simple formation, surrounded by police, a larger crowd surrounding them, and the White House in the near distance. I looked from beyond vision, from beyond the space we were in. I whispered to a friend on my right, "This is a dance, a ritual; we are playing parts in a much bigger scene ..." He found my hand and held it. The man on my left, someone I did not know, reached for the other hand, and we began to sing softly. The police gave their final warning. Hearts were beating; breath shortened; the earth pulled us to her from below. Gravity pulled firmly, gently, unmistakably, up through the pavement to the body. We were one.

The policeman was speaking again through the bullhorn. He flubbed his lines, grinned, and began again. We laughed. He laughed. They would not hurt us. We relaxed ... waited. The ball was in their court. Beyond the police on the street in front of us were other police officers in other parts of the city, directing and organizing, and beyond them judges, legislatures, and government officials. Everything was carefully coordinated ... choreographed ... from far beyond what we could see. Behind the law were municipalities, shops, businesses, schools, families, factories, hospitals, each wheel turning the next. Each turns on principles of order, equality, law, democracy, property, economy, security, civil rights, and civil society. Each turns by virtue of the others, providing, protecting, maintaining.

We, sitting on the pavement, now, were at the edge of the system—outside the system, outside the law, slowing the wheels, gumming the works, standing in the path of forward motion. We offered disruption, noncooperation,

Pipeline protest at the White House, August 25, 2011. (Ren Schild/Shutterstock)

inconvenience, and distraction. Our wheel did not turn with the others. There was danger here of breakage, of gears stripping and teeth gnashing. There was danger of disorder and the downfall of civilization. We were tense, anxious, and fearful, but polite. We asked the police to dance, and they accepted. Our hands were held to them, surrendering motion to their lead. They took us in their arms. The earth pulled us to her bosom as we rose with her, embracing the system, forcing it to dance, to look into our eyes, at her.

The music started as they picked us off, one by one. The day passed from cool morning to hot afternoon. As each of us stood, a team of muscular young men moved in and tied our wrists behind our backs with a black plastic strip. We could no longer reach for keys, brush away flies, scratch itchy noses, or use the toilet. There was no way to sit comfortably. The police led me to a booth where my hat was pulled off, my hair flew up in a swirl, and they snapped a photo. I asked the officer if he could lend me a comb. He laughed and snapped another photo. They led me to a wagon waiting in the sun. There were benches along each side with a solid wall down the middle. The rear door opened; four men were already sitting, sweating, and I took the next to last space on the bench. My knees were an inch from the dividing wall, my face not much farther away; I leaned uncomfortably forward with my hands cramped behind my back against the outside wall. I was lightly dressed, but sweat beaded on my nose. The man next to me wore a business suit. We could hear muffled voices on the other side of the dividing wall. The air was heavy, hot, and motionless. We waited for nearly an hour.

We introduced ourselves. I couldn't see anyone else clearly, and I was separated from the friends I came with. Each voice was a blend of jubilation, pain, euphoria, and discomfort. No one was suffering seriously, but I worried about being so tightly packed and so helpless. What if there was an emergency—a fire, or an accident? We would not be able to get out. But we were without choices. We were tight in the arms of the law and would follow the music until it stopped. Saying very little, knowing less, a bond of fellowship rose among us as we sat, stifling, waiting, the pain creeping up our arms and shoulders. We were brothers, suddenly, accidentally. We shared something in our minds and now something in the world. Each of us chose this, risked it, thought about it long and hard, gave time and money to do it. We said we would do it, told anyone who asked, and now, here we were, alone together. No one was here without conviction. No one told us to do this. We were convicts, convicted, with conviction. Someone said something about the "paddy wagon five" and wanted e-mail addresses, but there was no way to write. Finally, the rear door opened and our final comrade stepped in. A cheer broke out: now we were the paddy wagon six! I grabbed a quick breath of fresh air before the door slammed shut.

We rode the potholes across town to Anacostia. We were jostled around; there was no air. We began to tell of what brought us to this, of how each came to be on the bench we shared. There were no heroes, only ordinary people who got to live briefly beyond their own lives. We agreed that this was a beginning, that there would be more. No one said where it was going, but each of us knew it was going on from here. We would disappear into our own lives and reappear somewhere, sometime. We would do this again.

When we pulled into the station house, I craned my neck to the rear window and saw two women from our group on the sidewalk, already released—a good sign. (The women were cuffed and brought in separately.) The paddy wagon parked in the sun, and we waited again. Clothes were soaked and throats dry. My head was light and I wanted out, now. I no longer moved to the music, but the band played on. We waited, not knowing for what or for how long. Claustrophobia seeped into my brain, and I pushed the air from my lungs. A policeman appeared and opened the door. We stepped out slowly, straightening cramped backs and legs, sucking fresh air. We filed into the air-conditioned station house and waited in line. An officer asked if I had been arrested before: "Hitchhiking, 1968," I responded with a slight grin, and he waved me on.

We could pay a fine or go to jail. The handcuffs were cut and thrown in the trash. Someone wondered if they are recycled. My wrists were marked and shoulders sore as my hands stretched to the ceiling and to the floor, relieved. I reached into my pocket for my driver's license and a one hundred dollar bill. To avoid booking delays, before we left this morning we had emptied our pockets of everything else: wallet, phone, keys, and pocket change. I qualified for "catch and release"—they took my money and let me go. They let me go! The DC police have done this dance before. Unlike some others, they know how and when to be gracious. It will not always be this easy.

They bowed, hand outstretched, bidding me adieu. I walked back across the floor.

The Paradigm

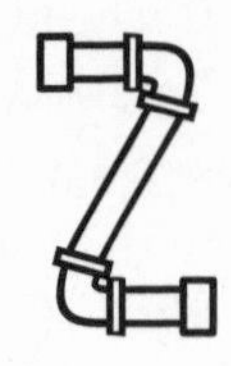

IF YOU GO OUTSIDE ON A CLEAR NIGHT AND LOOK UP AT THE stars, you will see the same sky the Babylonians saw four thousand years ago. Stars, planets, and moon all rise in the east and pass slowly overhead as the night progresses. The ancients thought the sky was a giant pinwheel centered on the North Star, turning slowly above the earth. But you know the sky doesn't actually rotate; it is the earth that rotates. You know this is what you are seeing, but what you are looking at is *exactly the same* as what the Babylonians were looking at. There is no difference in the sky.

You know more than your eyes are seeing. Or, more properly, *we* know more than *our* eyes are seeing. You didn't figure out by yourself that the earth revolves around the sun; you learned it in school. You know what you know because you are part of a collective consciousness. As modern people, we have a different way of seeing the same sky the Babylonians saw. We know more than they knew because something major has happened to who we are. We are no longer individual creatures crawling across the earth's surface looking at the sky as individuals. We look at it together, and think together, and find ways to check what we are thinking to see if it makes sense. That is why we perceive something so radically different when looking at the same sky. The earth-centered and sun-centered understandings of the solar system are not just differences of opinion; they are differences in who we have come to be and how we understand ourselves. These are differences in *being*. The movement from the geocentric to the heliocentric solar system is a *paradigm shift*.

The term was popularized by Thomas Kuhn in his 1962 book, *The Structure of Scientific Revolutions*.[4] The *paradigm*, according to Kuhn, is not a theory or opinion, but the worldview within which theories or opinions are held. It is the difference between preference and undeniability: you may prefer to believe that the universe rotates around the earth, but you cannot reasonably deny the evidence that it does not. Scientific revolutions happen, Kuhn explains, after a series of anomalies are discovered that cannot be explained by existing assumptions. Understanding is hurled into a crisis, and new ideas that are plainly absurd according to the old assumptions have to be considered. (The earth cannot possibly move—all the buildings will fall down!) A new paradigm takes shape and develops a following. But there is little or no middle ground between the old worldview and the new; each clearly contradicts the other at some critical point, and all-out intellectual warfare ensues. The shift between paradigms is not gradual or evolutionary, Kuhn says, but a "series of peaceful interludes punctuated by intellectually violent revolutions" until "one conceptual worldview is replaced by another."

Kuhn's book was a paradigm shift of its own. It revolutionized understanding of how science leaps from one worldview to another and how human understanding as a whole progresses in discontinuous fits and starts as new information accumulates. The book was, in fact, a little too successful for its own good. The concept of the paradigm shift became so popular, and so popularized, that it seeped into everyday life and language, even into advertising, and lost much of its original impact. We seem now to experience paradigm shifts every time we paint the kitchen or choose a new brand of toothpaste. There's even a *New Paradigm* engineering company busily designing ways to suck tar sands out of Alberta. But the term retains the meaning Kuhn gave it. He specifically restricted its use in his own work to the nature of progress within scientific communities, and defined it as a fundamental change from one worldview to another—the difference between what a Babylonian shepherd saw in the night sky and what you think you are looking at now.

It is also the difference between a fragile deciduous forest and so much overburden. Both look the same, but they exist in entirely separate worldviews. Or the difference between caribou habitat and a billion dollars of raw tar; both occupy the same place in physical space, but each is a distinct, contradictory reality. They cannot exist simultaneously. Either the forest and the habitat become overburden and tar or they stay forest and habitat. Either the *economic* paradigm prevails, or we shift to the *ecologic* paradigm. There is precious little ground in between: the earth is not both alive and dead at the same time.

The lack of middle ground marks the transition as a paradigm shift and not an evolutionary process. The leap from one worldview to the other is thorough and decisive. The shift is often compared to a Gestalt: when shown a drawing of two faces looking toward each other in profile, the mind will shift at some point to seeing a vase. You can see the vase or the two faces, but only one or the other; you cannot see both at the same time. Opinions linger in the gap between worldviews, but worldviews go one way or the other.

The transition period between paradigms is often characterized by a frantic effort to shoehorn new observations into the old paradigm. As evidence of a new paradigm becomes undeniable, creative compromises spring up in an effort to preserve the old worldview. Tycho Brahe, for example, the leading astronomer in the generation after Copernicus, came up with the perfect compromise between the earth-centered and sun-centered views of the universe. His "Tychonic" system allowed the planets to orbit the sun, as did the Copernican system, but kept the sun *orbiting the earth*. This explained retrograde motion of the outer planets just as well as the Copernican system did, but it preserved the old paradigm with an immobile earth at the center of the universe. It kept astronomers happy by explaining planetary observations while keeping non-astronomers happy with the continued assurance of humanity at the centerpiece of God's creation. The universe would continue to revolve around us, and we would not have to change who we thought we were. But the Tychonic system was a reassurance and not a revolution. It did not become a new paradigm because it remained entirely within the old paradigm. You can still "see" it in the sky on a clear night, but it did not stand up to Newton's theory of universal gravitation. It could not explain *why* the planets move the way they do and collapsed as further evidence accumulated.

As the ecologic paradigm challenges the economic, attitudes will cluster around a supposed middle ground. Compromises will arise between extreme positions on either side as reasonable people try to work out a balance between energy use and environmental preservation. There is a wide range of opinions, for instance, within the "mixed use" idea, where wild areas can be exploited for both resources and recreation. Many are convinced that it is possible to "reclaim" abandoned mines and restore them to as good or better condition than before extraction by building recreation facilities, planting trees and grasses, and introducing attractive megafauna after the damage has been done. Old mines become available for new uses: tourism, recreation, hunting, hiking, and boating. These become possible where they did not exist before. But the concept of "use" reflects the old paradigm; land is understood only from the human perspective, as a resource. This is an extremely limited understanding of the natural world and of the larger universe of which we are a part.

In the ecologic worldview, biological communities exist in their own right, whether or not they are "useful." Trees, soil, animals, streams, plants, lakes, deserts, and oceans have no less right to exist than human communities. As a living community ourselves, we are in the same overall picture as they. Recreation, conceived as an economic activity, is not what the land is *for.* Wilderness areas should be preserved even if they do not generate income for anyone.

Within the ecologic paradigm, all forms of economic activity, including recreation, are part of the living process and have a rightful place in relation to other living processes. The living world *includes* the economy, but it is not limited to the economy.

The old paradigm sees the earth through the lens of money. The earth is raw material, a component of economic growth. Economic growth is the reason to do things. Business and government are organized for the purpose of making profits by meeting consumer demand. We hear over and over of "the growing need" for more electricity, bigger houses, more cars, and more roads to drive them on, and we are led to assume that it is the mission of society to fill that need. We set "needs" first, and then set our relation to the earth within that limitation. We want our fill of carbon-based fuel before beginning to consider the possibility of ecological imbalance.

In the new paradigm, human need is subordinate to the earth's capacity to sustain life. The earth is before, after, and on all sides of us; it is bigger than us; it envelops us. We fit into *it*. Like every other form of life, we are an expression of the earth, and we build our economy within the living world. The economy is a wholly owned subsidiary of the ecology.

Yet it is hard to see that truth directly. The economy has its stranglehold on the human mind, body, and spirit, and it will not yield easily to a clear view of the human condition. A few well-known but poorly understood facts illustrate that truth. We live in a time and place (early twenty-first century America) of unprecedented prosperity. There is more wealth—far more wealth—than has ever existed at any time or place in human history. We have more stuff than anyone, anywhere, has ever had. Yet it is not enough. It is not nearly enough, and the thrust of our collective efforts is pointed at producing more. Acquiring more money and more stuff will always be the central purpose of social being within the economic paradigm.

Our political issues revolve around the need for businesses, social classes, and interest groups to have greater economic power than they currently have. It is true that whole communities suffer without enough money to pay for houses, cars, health care, and education. The suffering is real—I do not wish to demean it—but most of the suffering in this part of the world is not poverty in the absolute sense, but *relative* poverty: the inability to live up to the material standards of the prevailing culture. That standard rises every year. Within my lifetime, the size of the average American house has doubled while the number of people living in it has halved.

I am speaking here of averages, not individual cases—social climate, not social weather. Unemployment, student debt, underwater mortgages, and health care catastrophes remain widespread and overwhelming. But *average* incomes and property holdings are higher than ever before. Or, more properly stated, income *demand* is higher than ever. Families can't "get by" with a single income any more. Even as income grows to higher and higher levels, spending increases more rapidly. The rate of saving has plummeted to near zero and at times reached into negative territory. Everyone needs more money than they have. One might think that people would reach a level of comfort: a nice house, good food, a car, education, health care, a steady job, etc., and things would level off. But the economic paradigm does not work that way. There is no such thing as enough. It is not about money; it is about *more* money.

THE CONFLICT BETWEEN THE OLD AND THE NEW—BETWEEN the economic and ecologic paradigms—is not a contest of logic or reason. There are always reasons for more money and reasons to protect the environment. Each paradigm is a complete system of logic and reason within its own assumptions. Each is a separate worldview and a separate world. To understand the logic of either, one must step within the perimeter it defines, relinquishing all allegiance to the other. Each is entirely "right" from within and "wrong" from without. Neither is irrational or unscientific. Neither is wrong in any absolute sense. The conflict can be civil and respectful, but neither side can call upon a universal standard or objective authority for the legitimacy of its argument.

The conflict is not between truth and fiction; it is between underlying assumptions upon which truth or fiction is constructed. The assumptions are more often unconscious than conscious, and the paradigms they shape differ not in what one is looking at, but where one is looking from.

Our present day understanding of the paradigm shift between the earth-centered and sun-centered universes, for instance, is conditioned by our seeing the world from the point of view of the sun-centered universe. We think of ours as the "real" or "true" universe, and the medieval, earth-centered universe as "ignorant" and "false." We tend to think of the transition from one to the other as an inevitable realization of what is really out there. But this does not take into account that the medieval world was complete and self-contained from its own point of view. It had everything it needed. The modern world forced its way into an already complete picture. From the standpoint of the old paradigm, the new paradigm is always an unnecessary intrusion.

Paradigm shifts are wrenching and painful not because ignorant people cling to old and outdated ideas, but because knowledgeable people stand up to defend time-tested truths against unwarranted, radical innovations. The economic paradigm is a complete and self-contained worldview. It does not need the ecologic paradigm and would like nothing more than to continue on its own without the inconveniences of living within the natural world, just as the medieval worldview would have liked nothing more than to continue happily on its own without Copernicus and modern science.

Much of what happens within the old paradigm continues in the new, though the purpose and meaning change. It is possible, for instance, to continue extracting and burning carbon *within* the ecologic paradigm. Extraction will not be of every bit of carbon available to see how much money can be made; it will be of the carbon that the earth's atmosphere can tolerate. Ecological balance does not mean complete cessation of mining and fossil fuel use. It means finding what the sustainable level of atmospheric carbon is and establishing ways to maintain it. It means the possibility of some fossil fuel use, if that is appropriate, especially during the transitional stage to renewable energy.

Under the new paradigm, the primary goal of human civilization will be the sustenance and nurture of life. A symbiotic middle ground will be found between human and non-human needs, but it will not be found from the standpoint of money; it will be found within the carrying capacity of the earth.

Will we all have to agree to make this happen?

No. Paradigm shifts are rarely accomplished through reconciliation. They happen slowly, over time, through gradual acceptance of the new paradigm by the younger generation. As naturalist Charles Darwin wrote in *On the Origin of Species,* "Although I am fully convinced of the truth of the views given in this volume ... , I by no means expect to convince experienced naturalists whose minds are stocked with a multitude of facts all viewed, during a long course of years, from a point of view directly opposite to mine, ... But I look with confidence to the future,—to young and rising naturalists, who will be able to view both sides of the question with impartiality."[5] Or, as physicist Max Planck once said, "A new scientific truth does not triumph by convincing its opponents and making them see the light, but rather because its opponents eventually die, and a new generation grows up that is familiar with it."[6]

Our grandchildren will one day assume as obvious what we fight for all the world to see.

February 20, 2012 – Louisville, KY
High 50°F – Low 24°F – Precipitation 0.0 inches

The price of gas is up. Mitt Romney said today that the Keystone XL pipeline is a "no brainer." It is.

Sun Power

3

February 28, 2012 – Frankfort, KY
High 65°F – Low 31°F – Precipitation 0.0 inches

I spent most of the day in Frankfort again, lobbying with sixty other people at the state capitol for the Clean Energy Opportunity Act. If it becomes law, the bill will create Renewable Energy Portfolio Standards (REPS) for Kentucky, allow the feed-in tariff, and provide incentives for energy conservation.

AS A SOLAR INSTALLER, I WAS INTERESTED IN ALL THREE PARTS of the bill: Renewable Energy Portfolio Standards, the feed-in tariff, and conservation. REPS would require electric utilities to produce a minimum percentage of their total energy from renewable sources. Companies who do not produce their own wind, hydro, or solar energy would have to buy renewable energy credits from people who do—people like my customers. This creates an income stream for them, and a major incentive for new customers to go solar.

The feed-in tariff would be even better. We now have something called *net metering,* which allows anybody to put solar panels on the roof of his own house for his own use, and remain attached to the utility company. The grid becomes a sort of electrical storage system. While the sun is shining, solar panels pump excess energy not being used at the house into the grid, turning the meter *backward*. At night, or when the sun is not shining, the house draws energy out of the grid as it normally would, turning the meter forward again. The amount paid to the utility at the end of the month is the *net usage,* or whatever amount the customer ends up consuming above what he produced. But the feed-in tariff would be a big step beyond this, because it would allow installations to produce *more* than they consume, with the excess sold to the utility. Installations, large and small, would be designed to produce solar energy, at a price, for use anywhere on a utility's grid system. This would provide all kinds of new investment opportunities using parking lots, empty fields, and warehouse roofs. People who want to go solar but who have too much shade or too many obstructions on their own roof could install a system somewhere else. My business would soar.

Interestingly, the legislature is much more interested in the business end of solar than in renewable energy itself, so we are promoting the Clean Energy Opportunity Act not as a way to cut carbon emissions, but as a way to create new businesses and jobs. If we convince them that we would make a lot of money and hire a lot of new people, they might pass something like this. A recent study showed that if this bill becomes law, it will create 28,000 new jobs in Kentucky.

And then there is conservation, the third component of the bill. Conservation, by itself, can be really boring. It means having less, doing without, having a colder house—or a hotter one. It means not having all the energy you might be able to use. It's all yin and no yang. But coupled with a set of solar panels on your roof, conservation becomes a power in itself. You feel the energy coming into your house, and you feel it going out: the yang mingles with the yin. You *think* about it; it becomes part of your consciousness, part of your day. You turn the lights off and set the water heater down a few degrees not out of eco-guilt or to save a few pennies, but because you saw the sun out and you know it fell on your roof, and that your meter turned back a full 16 kilowatt-hours today. You don't want to watch all that good sunlight running down the drain. You've got a well designed system, perfectly balanced to produce all the energy you really need, but to keep it balanced you have to be aware of what the sun is doing and what you are doing. You do the balancing, in accordance with nature.

You stop thinking of energy as so many dollars and cents. Your energy use becomes *demonetized*: energy is not money but sunlight! It's free! Like a tree, you have spread your foliage into the light. But you have to use it well. You feel energy flowing through the roof, out through the wall sockets, and you want to take good care of where it goes from there. And, there's plenty of it! Your house becomes a living system: a microcosm of the biosphere itself.

I have noticed this change in consciousness in my own case, and in *every one* of my solar customers. They *all* end up using less than they did before, and not because they like to "shiver in the dark," as President Ronald Reagan would say, but because they have come to live with energy as a life force rather than a commodity.

Not all of my *prospective* customers are like this, however. I get calls all the time from people with huge houses and even bigger energy bills. They are using so much electricity that either I would have to install an $80,000 system that they don't have the roof space for, or I could install a reasonably priced system that will only offset one-fourth of their usage. It's all yang and no yin. I have never refused to install for such a customer, but I invariably end up talking them out of it. They are all fine people, but it's the old paradigm: lifestyle comes first, and the environment fits into the picture somewhere else if it's cost effective. People who are using four or five times as much energy as the national household average generally have no idea how much they are using and have enough money not to think about it. They have no incentive to use less. Some of them want to use less, but tragically, they cannot. Waste is literally built into their houses, in their furnaces, air conditioners, pools, pumps, hot tubs, hair dryers, electronics, etc. There is very little they can do, or that I can do for them. I often suggest, halfheartedly, that they put in a moderate sized system, but if they are using four times as much as it puts out, what's the point? Solar *is* a good, profitable, long-term investment, with a payback period of about ten to twelve years, but if you're only in it for the money, you can find a better investment. Hey, if you're just looking for a place to park some cash, buy a savings bond!

March 1, 2012 – Louisville, KY
High 62°F – Low 44°F – Precipitation 0.0 inches

TransCanada, the owner of the Keystone pipeline, has announced it will build the southern leg through Oklahoma and Texas without presidential approval of the project as a whole. With state governments allowing the power of eminent domain to be exercised in some cases, pipeline construction will begin soon between Cushing, Oklahoma, and the Gulf of Mexico. The president, despite overtly opposing the pipeline, seems to be encouraging TransCanada to keep pushing it. He applauded the go-ahead on the southern leg and said he would "take every step possible to expedite the necessary federal permits."[7] Maybe his disapproval of the pipeline as a whole is just a temporary appeasement of a few environmental wackos after all.

March 5, 2012 – Louisville, KY
High 37°F – Low 28°F – Precipitation 0.29 inches

I am planning to build a carport. I have cut cedar poles and completed the foundation work already. I need lumber, metal roofing, and help. When the time comes, I will buy a case of beer and get a bunch of friends over here to raise the poles and attach stringers, rafters, roof sheeting, and hopefully the roofing itself. With good weather, and not too much beer, we should be able to get that much done in a day.

I HAVE NEVER NEEDED A CARPORT BEFORE. IT SEEMS A LOT OF expense to keep frost off the windshield. But I'm not building this one for the car; I'm building it to fuel the car. I figure that a single 235-watt solar panel will provide roughly enough energy to drive an electric car a thousand miles per year. That's incredibly efficient, way more efficient than a gasoline engine. Electric motors just run more easily than internal combustion engines. They are quieter and a lot easier to maintain, or so I hear. Electric cars, besides ruining the petroleum industry, would ruin the parts distribution industry: no more carburetors, spark plugs, fuel pumps, timing belts, or exhaust manifolds. An electric car works a lot like an electric clock, and I want one.

But the electric car by itself is not the answer to the energy problem, much less to air pollution and climate change. Switching to electricity without switching to solar or wind or some other kind of renewable energy would be worse than staying with gasoline. An electric car would be a lot cheaper to operate than a gasoline car, but electricity is not an energy source; it's an energy *medium*. It moves energy from one place to another—in this case, from the power plant to the wheels of your car. The electricity has to be generated from an energy source, be that coal, natural gas, wind, solar, or nuclear.

About half the electricity in the United States comes from coal, which produces more carbon and more air pollution than petroleum. Burning less gasoline and more coal would not help the environment at all. But converting to electric cars would be a giant step toward getting away from either of them. I'm planning to put twelve panels on my carport: twelve thousand miles a year without carbon, mercury, sulfur dioxide, tar sands, oil spills, or wars in the Middle East. Like any solar installation, the cost is all up front, but once it's there I'll never have to buy a gallon of gas again.

Or almost never. An all-electric car would be great for driving around town, but I'm out on the highway too much. When the battery gets low, there's no refill at the next exit. When my battery gets low, I would be stuck along the side of the road until ... what? Somebody stretches an extension cord from the nearest outlet? Even if I catch the battery level before it gets too low, there's no way to recharge it unless I can find somewhere to plug in, and then I'll have to wait five hours for a charge. So, I don't want an all-electric car; what I want is a plug-in hybrid. I will plug it in every night and drive the next day until the battery runs down, and then the car will switch over to the gas engine automatically. It's an electric car with a gas backup. It's not a standard hybrid like a Prius—that's a gasoline car with a battery backup. Standard hybrids are more efficient than most cars, but their only fuel is gasoline. The gasoline engine generates the electricity used by the electric engine. To make full use of my solar carport, I need a vehicle that uses all electricity almost all of the time.

But there isn't one yet. Plug-in hybrids are just now coming on the market, and right now their best all-electric range is only thirty to fifty miles. I need a hundred. So I'll just use the new carport to keep snow off my car until the free market comes up with what I need. And it will, if enough people demand it.

SO ... IF EVERYBODY BUILDS A CARPORT, INSTALLS SOLAR PANELS, and drives an electric car, will solar energy be enough to meet all our energy needs in the future?

No.

The question itself presupposes the economic paradigm. It assumes that the conversation begins with "needs." Even with a lot of new panels on rooftops, we will not be able to pull the plug out of the coal mine and plug it into the sun. It won't work that way. Solar will not be a "solution" or part of "the overall solution" to our energy "problem."

The question should be, "How much energy can we produce from renewable sources, and how can we live within a realistic energy budget?" In time, there will be plenty, but we have no right to demand more than the earth can sustainably provide. The question should begin with the carrying capacity of the life system we are within.

As things now stand, the productive capacity of solar is limited. The best crystalline solar panels currently available are around 16 to 18 percent efficient, and amorphous solar materials are about half as efficient as that. Most panels are rigidly mounted with a set orientation toward the sun and produce near capacity only during the middle of the day. They produce very little in the morning and evening hours, and nothing at night. They produce under cloud cover, but at reduced rates. The atmosphere reduces solar energy on the earth to around one-fourth of what it is above the atmosphere. There will always be earth-based solar to one extent or another, but to go solar in a big way will require large-scale solar arrays in space.

That's not a big problem. It's a big project—a huge project—but it's not a big problem. We know how to do it. The modern solar energy industry began in space. Photovoltaic panels were invented by American engineers as part of the earliest orbital missions. Batteries, fuel, and generators are heavy and prohibitively expensive to lift into orbit, and there is a lot of sun power up there. Panels in low-earth orbit are above clouds and atmosphere. Far enough away from the earth that there is no night, they can be oriented directly toward the sun at all times. In the absence of wind and gravity, large arrays can be constructed with

The carport under construction, without solar panels or a 100-mile plug-in hybrid.

relatively little supporting structure. There would be no problem collecting all the solar energy anyone could ever think of using. The problem would be getting it down to the earth safely, but there are a lot of possibilities for solving that. I am not in a position to choose one of them or some combination of them, but I do not think this type of project would be beyond human capability. It would require an enormous capital investment and a thorough understanding of how it would affect the biosphere.

Life on Earth depends on the sun. We all know that. But life did not evolve in sunlight from the beginning. The earliest forms of life were *chemosynthetic* rather than *photosynthetic*; they got their energy—unsustainably—from chemical compounds floating about in the primordial soup. Photosynthesis evolved later, in a single-celled bacterium, probably in shallow ocean waters or tidal pools. Energy came from *outside* of the primordial soup, outside of the living world. But photosynthesis was not efficient in its original form: seawater blocked most of the sunlight. Each cell had to maintain direct exposure to sunlight; it had to be close to the surface of the water and free of intervening sediments, debris, or other photosynthesizing cells. More efficient means of collecting sunlight evolved on land in the form of the plant kingdom. Complex cell communities became multicellular organisms with specialized photosynthesizing cells placed high enough above the ground to maximize solar intake.

Plants are essentially living solar installations; roots, stems, branches, and leaves feed, nurture, and support tiny panels that reach up from the earth toward the sun. They gather energy and store it in the form of chemical bonds between carbon and hydrogen atoms. Coal, petroleum, and natural gas are long-dead plant materials high in hydrocarbon energy that we gather for our own use now. Fossil fuels are, therefore, *indirect* forms of solar energy; the energy you use to light your house or drive your car *comes from the sun*, but it must first pass through hydrocarbon bonds stored for millions of years *within the biosphere*. Energy from long ago has to pass through molecules in plants to become useful to humanity.

This is the problem. The release of unbalanced biochemical energy is disturbing atmospheric chemistry and disrupting the climate. In the very near future, we will learn to photosynthesize our energy from outside of the biosphere—directly from the sun—and store it in ways that do not destroy natural balances. Sustainable energy will no longer come from within the postmordial soup.

The life force behind plant evolution is now behind ours, speaking to us in languages we have not yet learned. As we listen, we will spread our foliage high above the earth, funneling energy through the roots and branches of a sustainable society. The Plant Kingdom will show us how.

A Look at Life

LIFE IS LOOKING.

March 6, 2012 – Louisville, KY
High 70°F – Low 29°F – Precipitation 0.0 inches

About seventy-five Lakota from the Pine Ridge reservation in South Dakota blocked two TransCanada pipeline trucks trying to cross their land yesterday. The standoff lasted for six hours, and five Lakota were arrested. Debra White Plume, one of the activists, reported, "The tribal police had to let the trucks get off the rez. They escorted them to the reservation line. We oppose the tar sands oil mine in solidarity with Mother Earth and our First Nation allies."

WHY AM I AGAINST THE PIPELINE?

I should be for it; I drive a car. Until my hundred-mile plug-in comes along, I'll be using gas like everyone else, and like everyone else, I would prefer to pay less at the pump. I even drive my car to go fight petroleum pipelines. When a dozen of us in three cars drove to Washington, D.C., last August to protest against the Keystone XL pipeline, we filled our tanks several times to get there and back, showing just how stuck we are on fossil fuel to do anything at all. We told ourselves we needed to fight fire with fire, but that's a rationalization to assuage a blatant contradiction. We are just as addicted to oil as everyone else. I'm no puritan.

Perhaps I am against the pipeline because it is easy for me. I'm not an oil worker, a pipefitter, or a coal miner, and I don't own stock in the petroleum business. I can stand at a distance and complain without suffering any consequences (other than getting arrested).

I could make up a set of reasons for my opposition to the Keystone XL, but I don't think reasons are really why people do things or think things. Reasons have a way of popping up on their own after things have been thought and done. I'm not even against the pipeline because of the pipeline. If there were a real way for humans to inhabit this part of the universe and still do things like the Keystone pipeline, I would be all for it. I would rather not oppose it. I would rather not stand in the way of those who will get jobs for building and maintaining it, or of those who will get checks in the mail for owning it, or of those of us (all of us) who would pay less for gas if it is built.[8] There are some wonderful life-affirming reasons to build it, and I don't have better reasons not to build it. I have reasons, but they are not better than providing a livelihood for a family.

So I look for where my reasons come from. I look at breathing and at gravity, and I feel the blood moving through my veins. Every day. I take time during the day to look at being alive. I don't see anything different from what anyone else sees, but I take the time to look. I watch what I am feeling and seeing and thinking. I watch myself watching. Some people call it meditation, but I would do it if it were called something else. *Meditation* has other meanings and thousands of years of traditions in cultures other than my own. To me it means looking at life. I'm not a Buddhist or a Hindu; I just look at being alive. Every day I sit and look. It's a habit I have.

The first thing I see when I close my eyes is English. I see a stream of images, sensations, memories, and thought fragments coagulating into thought patterns in my native language. I put things together in words and think of things I could have said, or should have said, or need to say tomorrow. There is a putting together of pieces into shapes that could also be words—a voice that is me speaking to me. This voice, with some editing, becomes the outer voice I use when speaking to others. Most people know me as an English speaker, so this is a big part of who I am. I have a special relationship to people who speak and understand this language.

After a few minutes sitting and watching, the English becomes softer. It's always there to some extent, blabbering away while I am trying to see past it, but it gets quieter and gradually relaxes. Thinking remains, but less of it is in English. I think of *things*, but I don't have words for them. The fire of thought flickers deep in my mind as I catch little images of places and colors and faces and shapes and lines, and of things there are no words for. I'm pretty sure everybody sees this all the time, whether or not they pay any attention. I don't know if birds and squirrels and fish and houseflies experience this, but I'm pretty sure all humans do.

I think this is the part of me that makes me a person. The English makes me American; the thinking makes me human. And I have come to understand, over the years, that this is not just a realm of consciousness that I *have*; it is humanity itself. It is a single thing. When my eyes are open it looks like there is one person here and another there; I am over this way, you over there, and we are separated by the space between us. It looks like we each see and think separately. That's the way it looks, and the way we think of it. But I believe humanity is thought itself. When I look into this realm of consciousness, I am not looking at myself, the human being, I am looking at being human. Even the space we see between us is part of the looking.

And it goes on from there. Being human is a major part of what I see, but below thought there are feelings and sensations that are not strictly human. They are not particular to human experience. There is nothing unusual here, nothing anyone can't see or feel at any time, but I'm quite certain you don't have to be human to see or feel it. There is the beating of the heart and the flow of blood through the veins and arteries. They are always there. The heart beats thousands of times each day, but you don't have to be human to experience it. If you look at the *thump, thump, thump*, you may be thinking "This is *my* heart beating, in *my* chest," but it is the *experience* of the beat that you are looking at, not something "out there" in an external world. The experience is not unique to you. Watching the beat and feeling the flow is the same for fish and birds and squirrels, even houseflies, and for other people. It is what makes them animals. You don't have to imagine what it is like being an animal; all you have to do is look. You know what it is like to be a rabbit or a dog or a lizard, because you are one of them. Feeling the flow of blood in your body is your connection to the streams and lakes and oceans, where the animal kingdom evolved.

And, as you look at being alive, there is always the breath. Each breath arises: up and in. Then it turns and falls: out and down. There is a universe in every breath; all you have to do is look at it. If you think, "Oh, that's just a 'breath,' " you won't see it as it is: air flowing slowly in and out. Animals breathe, but so do plants. They move air in and pass it out, as you do. You breathe in; they breathe out. You breathe out; they breathe in. They breathe in the carbon dioxide you breathe out, and they breathe out the oxygen you breathe in. It is a great sharing. Breath is the experience of plant being. It is prairie and forest. It is the part of being alive that you hold in common with plants.

Below the motion of breath and flow of blood, there is the solidity of the body and the stillness of bones, joints, and muscles. There is the force of gravity pulling on your arms and legs, tying you to the rock and soil of the earth. Feeling the force of gravity is feeling your connection to the earth. It is the earth itself. It is the looking you share with the ground.

This is just a little something I do every day. There is nothing especially mystical about it. There is nothing objectively provable about it either. It is the experience of life.

It is why I am against the pipeline.

Civil Disobedience

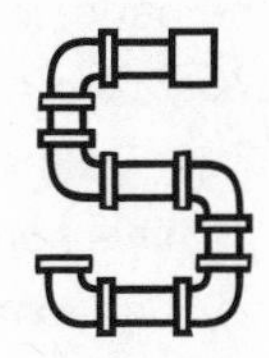

March 8, 2012 – Louisville, KY
High 64°F – Low 39°F – Precipitation 0.69 inches

The Senate just voted the pipeline down! It was close: fifty-six voted in favor—more than a majority—but they needed sixty. The measure came up as an amendment to the transportation bill. President Obama had to lobby Democratic senators to be sure it would be voted down. That's three times the pipeline has been defeated, and not the last time, I am sure. TransCanada has already said it will reapply. There is too much momentum and too much money behind all that carbon fuel.

One of my senators, Mitch McConnell, came out with a statement yesterday condemning the president for deliberately making gasoline prices high to promote alternative energy sources. Equating "energy *policy*" with maximizing energy *supply,* McConnell expressed impatience with the president's disapproval of the pipeline.

March 9, 2012 – Louisville, KY
High 55°F – Low 36°F – Precipitation 0.0 inches

A talk show on NPR discussed the Senate vote briefly this morning. The perspective was entirely political: all forty-seven Republicans plus eleven Democrats voted in favor. All the Republican presidential candidates are for the pipeline, and Obama, the pundits said, has to be against it because he ran as an environmentalist and has to do something to look like one. Potential jobs range somewhere between two thousand and twenty thousand (short-term construction jobs, for the most part), and there was mention that all of the oil would be sold overseas. (This is unlikely: much of it will be sold in the United States, depending on world market prices.) But what disturbs me is that there was no mention of carbon emissions, no mention of tar sands, and no mention of climate change. None. The environmental issue, for those on the panel, was entirely limited to avoiding a route through the Sand Hills of Nebraska. If the route is changed, which it likely will be, it seems that Obama will have the cover he needs to approve it. He can claim to be protecting the Ogallala aquifer in the Sand Hills against potential oil spills, rather than "appeasing extremists," those fringe environmentalists worried about the carbon bomb and environmental devastation in Alberta.

On the *PBS NewsHour* tonight, David Brooks called Obama's decision on the Keystone XL pipeline "insane." "It doesn't do anything for the economy," he said, "and it doesn't do anything for the environment."

IT IS CLEAR FROM LISTENING TO POLITICIANS AND PUNDITS that the ecologic paradigm is not well represented politically. It is present in all of us to an extent: we all know in the depths of our souls that human civilization depends on clean air, clean water, healthy oceans, healthy forests, healthy soil, sunshine, and a stable climate. We will decline and fall without them. But the depths of our souls do not put gas in the tank or food on the table. How much money things cost and how much of it we have in our pockets occupies far more of our waking consciousness, and most of our decisions are made, individually and collectively, at this level. We have to get through the day. It is at this level that we relate to one another for at least six days of the week. Money is something we all understand, something we all relate to in one way or another, and something that binds us into a collectivity. We may not worship the dollar, but we *believe* in it. We all think those little pieces of green paper in our pockets are real and have intrinsic value. If we did not—if we all woke up tomorrow morning and agreed that they were just paper—millions would starve.

Money is money because we believe it is money. The illusion of value becomes real when we hold it in common. It cannot be surprising, then, that much of what we decide to do collectively—in effect, politics—is based on something we believe in collectively. In a democracy, political representation is based predominantly on how individuals and interest groups relate to the exchange of money. The lion's share of collective thinking and collective action centers on how we bring money into the collectivity and how we send it back out.

The exception to this is national security. The ecologic paradigm will remain poorly represented politically until it is seen as a matter of security. Prosperity will be sacrificed for security, but only for security. Economic thinking will dominate the collective mind until the force of life rumbles deep within us—until our lives are threatened. It is then, and only then, that the paradigm will begin to shift.

But security will not be national security. National governments only provide security versus other nations; they do not provide environmental security. They have no relationship to the wholeness of the air, the water, the land, or the climate. They cannot speak for the wholeness of humanity or create policy vis-à-vis the non-human world. I will explore this aspect of the ecologic paradigm later on.

The ecologic paradigm, despite its current political weakness, is far more powerful than the economic because it is rooted in biological security. It does not deny the economic imperative—it is not fundamentally at odds with human activity within the biosphere—but it sets the value of *life*, human and otherwise, above economic growth. The ecologic paradigm finds a place for the human economy within the living world. *It understands economic forces as living forces.* Every form of life needs raw materials, energy, and proper waste disposal. Every form of life needs to grow, compete, disintegrate, and return to the soil. Every company, every enterprise, every shop and retail outlet draws nutrients from the land, seeks its place in the sun, finds its niche, competes with neighbors, feeds employees and customers, and in time closes its doors as the economic climate changes. The economy is a form of ecology that fits within a larger picture.

But there is a major difference between human beings and other species. The force of life flows through every organism in a linear pattern. Energy, water, and nutrients come in one end, combine and process within the body, and come out the other end as waste. Every organism of every species does the same thing in one way or another. Living *systems*, on the other hand, become nonlinear *ecologies*, in which one form of life feeds on the waste products of another. Plants exhale oxygen, animals excrete nitrates, leaves fall from trees and feed bacteria, and it is this synthesis of biological activity that creates the *balance* that life as a whole requires. An ocean, a desert, a forest, or a prairie is a living system that requires healthy interdependence among separate species.

Until the industrial revolution, humanity was a species with a linear flow of life. Now we are a living system, with the roots and leaves and branches of our economic activity reaching the limits of the earth's carrying capacity. We are an entirely new form of life with a new ecological base, and too big to not know or not care how we affect the rest of life. Before, we did not have to worry about the atmosphere, the soil, the oceans, or the forests; they took care of themselves. Humanity was too small to affect the planetary balance of nature. Now, *unlike any other species*, humanity has to understand itself as a whole; it has to take into account where it gets its energy and nutrients and where it puts its waste products. It has to fit its system in with other systems. The force of life flows through us not in a line, but in a circle. If the earth were infinitely large, we could feed forever into one end of the economy and dump out the other, but the earth is the size that it is and no larger. We have touched the edge of the biosphere and will have to close the loop to stay within it.

But how do we close the loop? If politics is so much about money, how do we curve the line into a circle? How do we translate into action the fundamental truth that economy is within ecology? It is one thing to sit on a hilltop and think wonderful thoughts about nature and clean water and flowers and butterflies, and quite another to develop renewable energy systems and manufacturing procedures that do not pollute lakes and streams. How will we ever develop the social machinery to balance our lives with the rest of the living world?

That depends on how smart we are. In a democracy, we get what we deserve. There is no guarantee of virtue or intelligence—wisdom has no more right to representation than stupidity. Good democracy gets people what they want, whether or not what they want is good. In theory, everyone is represented equally: we work out differences, compromise when we have to, and make collective decisions that become the basis of action. When the collective eye sees that ecology is a matter of security, the economy will become secondary; votes will go for renewable energy and cyclic resource use. It *will* happen: there will be a time when people—even the greediest among us—will vote for sustainable living. But when? Will it happen when the oceans are already dying and extreme weather is already ravaging the land? Assuming something like a democracy still exists, the real question becomes, how do we see what is coming before it comes? How does abstraction become reality in the collective mind before it shows up in the physical world? And will it be too late?

Will we wait until we can touch the tiger before we believe it is real? Or will the collective mind's eye see the stripes before the tiger leaps?

CLEARLY, THE ECOLOGIC PARADIGM LIVES IN SOME PEOPLE more than others, as does the economic. Some people already feel in their hearts the connection with nature, the oneness with all of life, while others are concerned more with keeping the economy going. The old and new paradigms do not divide us in any absolute sense; each exists to some extent within everyone. No one wants to kill nature, and everyone has to make a living. Each worldview is a collective vision that exists over and above the particular individuals it inhabits. But there are individuals within whom the new paradigm cries out most passionately, and it is these who will act to create new consciousness in the collective mind. It is these people who will show the rest of the world that ecology is a matter of security.

If something is true, it comes out in time. If we are more human than we are white or black, for instance, that truth will be revealed eventually. But it takes individuals who see that truth to break through the walls of ignorance that conceal it. When I was a child, I thought the civil rights movement was about "being nice to Negroes." I didn't know it was about being the *same thing* as Negroes. Some very dedicated individuals labored long and hard to work that into my head. I didn't see the truth at first; I had to be shown. It came to me eventually, not because their opinion was better than mine, but because what they stood for was a fundamental truth: people are people first and kinds of people second. Adolf Hitler, another very dedicated individual, tried to work the opposite idea into our heads a generation earlier, but what he stood for didn't take, because it wasn't true. What he stood for failed not because he lost the war; it failed because it was not true. It was hogwash. He could have won any number of wars, but his idea was doomed from the start because it was wrong.

If it is true that the ecology is larger than the economy, that truth will rise to the surface in time. We will realize it and wonder how we could ever have thought otherwise. But on its own, it will not rise to the surface *in time*. By the time there is good, hard, direct visual evidence that we have undermined the ecological basis of the economy, it will be too late to undo the damage. The tiger will have leapt.

Some say that even now it is too late; even with drastic measures to curb pollution and cut carbon emissions, too much damage has been done to reverse the downward slide. Even if we were to stop carbon combustion right now, our sins of emission in the past would be visited on our children for generations to come. We have tipped the balance—the point of irreversibility has passed. The past has already ruined the future. Those who say this might be right, but I prefer to believe that we have time to act. I do not know this is true, or know how much time we might have, but I am confident that the sooner we act the easier it will be to restore the balance.

I prefer to believe that we have time to act not because I know it is true, but because there is no point in believing otherwise. We cannot act in the past. We have to act now and in the near future.

But how to act? Those of us who see what we think to be true in the abstract must find tangible manifestation of that truth. We must find a symbol: something true in itself that represents a larger truth. We must place our bodies in the path of a small thing destroying a small part of the earth in a manner that symbolizes a larger thing destroying the earth itself. I am reminded of the lunch counter sit-ins in North Carolina in the early sixties. Young black people sat where they were not allowed to sit—not because they wanted to eat lunch, but because they wanted to reveal the injustice of racial segregation. A segregated lunch counter symbolizes a segregated society. They were polite, nonviolent, and passively disruptive. They showed people the truth that they lived every day, and in doing so created a new consciousness of what it means to be an American. They changed what it means to be a black person, and they changed what it means to be a white person. They changed me. Their message got through to the majority on the other side because they were respectful of the other side. We should learn from them to be quiet, strong, peaceful, polite, and nonviolent as we gently disrupt the economic paradigm at critical times and places.

Whether we actually stop or delay a particular project is much less important than how deeply our actions penetrate the inertia of popular opinion. We do not have to win all the battles or stop any particular project to win the war. And we must not be attached to our opinions, no matter how strongly we may hold them, as the truth we reveal may be something else. Respecting and honoring the opinions of those who oppose us will bring truth to the surface as much as anything else we do, because what they do will be as important as what we do. I learned as much from those who physically and verbally abused the lunch-counter protesters in North Carolina as I did from the protesters themselves. I like to think that many of the abusers may have been transfigured on the spot.

The Keystone XL pipeline is the fuse of the Alberta carbon bomb. It connects billion of tons of tar sand carbon to the atmosphere. In scope it is local, regional, national, transnational, and global. The pipeline is a real physical presence in itself and symbolic of a reality larger than itself. It will devastate the lives of indigenous people who live in and downstream from the tar sands region, endanger those who live along its route, and put at permanent risk those who live in the world it will forever alter. It will change life on Earth. A handful of forward thinking people will not stop it physically, but if they *try* to stop it, there is an excellent chance they will bring revolutionary new awareness to the society that witnesses them in action. The pipeline will become a symbol of environmental degradation as a whole.

NONVIOLENT CIVIL DISOBEDIENCE DOES NOT TELL ANYONE how to think or what to do. Instead it says, "Hey! You have to look at this! You have to look, whether it is convenient for you or not!" It seeks attention and points at a symbol of a larger reality. It is arrogant—at least as arrogant as it is heroic—but it is polite. It is aimed at people who are polite. It is aimed not at leaders, but at the public from which leadership emerges. It does not create policy; it creates the consciousness that is the basis of policy.

And let us not fall into the trap of blaming the ecological crisis on the people who provide us with fossil fuel. At this point, we need what they have to offer. The food we eat, the way we get to work, and the light and heat in our homes all come to us by way of their services. They sell only what we buy. When we go with our physical presence to protest fossil fuel, we need their fuel to get there. We should thank them for it. We should protest what they do, but we should not protest who they are.

There is a certain safety in this. We don't have to make them wrong to be right. If all we are doing is pointing at what they are doing, we don't have to justify ourselves by vilifying them. We don't have to tell people what to think—just point at the pipeline. Look at what is happening!

Showing—not telling—is the great accomplishment of civil disobedience movements in the past: young black people sitting quietly in the face of abuse at a lunch counter in Greensboro; demonstrators refusing to retaliate as Bull Connor trains fire hoses on the crowd in Birmingham; Gandhi's followers at the Dharasana Salt Works marching peacefully, row after row, as they are beaten to the ground. Actions such as these change how people understand the world and themselves.

Find a simple truth and bring it to the surface. Find a symbol. Find a project, a practice, a law that stands against a universal truth and resist it, peaceably. Be polite. Be gracious. Clean up after yourself. Thank people. Congratulate the police for a job well done. Apologize for disrupting the day. Disrupt it, then apologize. Help people to see things by being strong but gentle. If you are met with abuse, verbal or physical, try to take it graciously. The abuse itself is important. Accept the risk before you go in. Be bold; be smart; be in control of yourself.

Violence does two things that nonviolence does not do: it attracts immediate attention, and it forces people to change what they are doing. Break windows, grab a gun, beat a few people up, and you get attention. Threaten them with injury and they will do what you say. Violence is effective in changing how people think and what they do, that is why it is so often used. But violence, besides hurting people, creates unity only by dividing. It heightens consciousness only by separating *us* from *them*. It drives a wedge between good guys and bad guys, left and right, demonstrators and police, countrymen and foreigners. It does not focus attention on creative solutions. It focuses instead on destroying the enemy in the belief that the problem will be gone when the bad people are gone.

Nonviolence works very differently. It is effective only when there is no enemy. There is always *opposition* in civil disobedience, but there's never an enemy.

Martin Luther King changed America not by defeating white people, but by showing them what an American is. Gandhi drove the British out of India not by pushing them out, but by refusing to be ruled. We will keep the earth habitable not by destroying oil companies, but by igniting the conscience of their customers. The opponent is always part of the show in any nonviolent campaign. We will depend on him. What he does—how he reacts—is as important as how we act. Help him. If the new paradigm is global, there can be no enemy.

The tradition of nonviolence began in the early twentieth century, but effective incidents of civil disobedience go back much further. Galileo, for instance, was arrested in 1633 for the heresy of promoting the sun-centered worldview. His bravery in standing up for what he knew to be true focused enormous public attention on the reaction of the Church to what he said and did. The Church was at the point of compromise; it was willing to accept the Copernican system as a possibility, but Galileo presented it as fact. In essence, he forced its hand. It was the Church's *defense of the old paradigm* that swept the medieval world away as much as it was Galileo's defense of the new.

Effective nonviolent actions in the modern world require organization, planning, and discipline. It is always best to attend an action with people you know and with whom you feel comfortable. Find a group of people, or form one, by going to meetings and rallies and hanging out with people you enjoy. You could start with a church group or a book club. Keep your group activist, but make it social as well. Party together. When you get to an action, the important thing is to know the people around you and know what to expect of them.

What has always worked for me is to keep the level of organization fairly low—e-mail lists, but no membership lists; contributions, but no dues; planning meetings, but no monthly meetings. This will vary from group to group, of course, but I have found that standing organizations take on a life of their own and require a lot of work to keep the program going, keep the members interested, and keep the dues coming in. They are absolutely essential as a structural backbone of any long-term movement, but you may not need *another* one for the specific purpose of civil disobedience. Meet when you need to, plan and take part in actions as they arise, then give it a rest for a while.

If your group is well organized—if you are trained, have what you need, communicate well, and take care of each other—you may be of special value at a large action that involves other groups. There may be a special task for you there. Having a number of preestablished groups at a large action is a tactical and logistical blessing for organizers of the event.

The nightmare of an event organizer is the spontaneous appearance of rogue individuals or groups bent on diverting the crowd's attention for their own purpose, or on confronting authority for their own reasons. One or two individuals throwing rocks, picking fights with counterdemonstrators, or assaulting police can ruin the message of a well-planned event. It happens regularly. Well-organized groups within a larger demonstration can spot people like this and intervene to defuse the situation. Often it is as easy as calming down an untrained individual and explaining to him the nonviolent spirit of the day, or it may mean assigning someone to isolate him, or report him to the police. If a group of people shows up intent on disruption, it is important that the police know who they are and that they are not part of your group or the action you have planned.

So come to an action with a group of friends if you can. Or make some quickly when you get there. You will gain the confidence and courage you need from these people. It is your solidarity with them that will change the world.

Finally, if you participate in nonviolent civil disobedience, keep in mind that you are helping develop a new form of social evolution that has been around for only a hundred years. Before then, nearly every major social change required violence. It took violence to expel the Persians from Greece, to create independence for America, to abolish slavery, and to defeat Hitler. Even since the dawn of nonviolent civil disobedience in the early 1900s, most social changes have been violent. Many people assume now that it will take organized violence to create major changes in the future. But that is the difference between the past and the future—violence will not do what needs to be done. We can no longer accomplish what needs to be done by dividing humanity into good guys and bad guys.

There is an undiscovered universe of new techniques for creating awareness through nonviolent action: how to choose symbols, how to train and motivate people, how to find the best times and places, how to work with police, and how to use new media. The technology is new, the spirit is new, and the issues are new. None of what we are about to do has been done before. We need creative people to develop new skills and new strategies for directing collective attention to what is happening in the world. We need people who understand the new paradigm to bring it to others.

We will use force. Not physical force, truth force: *Satyagraha,* to use Gandhi's term. Bringing truth to the surface, whatever it may be, requires peaceful, polite, gently disruptive force: a new variety of social motion that we are only beginning to develop. It does not have a long tradition. We will make it longer.

Fight politely, fight without fighting, fight as if your granddaughter's life depends on it.

AT THE PRICE I AM CURRENTLY BIDDING LARGER INSTALLATIONS, the same $40 billion, were it invested in solar energy, could provide sixty billion electric car miles every year for the next forty to fifty years. If there are 250 million cars in North America, that would provide somewhere around 240 miles per car per year—not nearly enough. But that's just the cost of the pipeline, not the fuel.

I don't need that much work anyway.

March 14, 2012 – Louisville, KY
High 82°F – Low 53°F – Precipitation 0.0 inches

I submitted a proposal this morning for a 53,820-watt ground-mounted solar installation at a substation owned by an electric cooperative. If the job goes the way I hope it goes, this would be a big step forward for me and for the solar industry in Kentucky. It will be a generating facility rather than a net-metering facility.

NET METERING, IF YOU REMEMBER, USES SOLAR PANELS TO reverse the flow of electricity through a particular electric meter. It is a way for a homeowner or a business owner to offset his own electrical use. But net metering is limited to the consumption at a particular metering site; that is, if the building where the panels are installed uses one thousand kilowatt-hours per month (about the average for a household), panels should be limited to produce only that much in order to create a balance between production and consumption. If they produce more than the building uses, the owner of the building is not paid the difference. There is no financial incentive, therefore, to produce more energy than is directly consumed at that site. This keeps the solar industry small. We are grateful to have net-metering legislation here in Kentucky, but we want further legislation that allows larger installations to generate electricity for sale to the grid as a whole. The installation I proposed today will help us promote the feed-in tariff to other co-ops and to the legislature. A living example speaks more loudly than facts and figures.

I also heard today that the Kentucky House Committee on Tourism Development and Energy has finally granted a hearing on the Clean Energy Opportunity Act that we have been lobbying for. Somebody said the right thing somewhere. The committee will listen to us harping on about the wonderful new world of renewable energy ... actually, they won't. We know better than to present a new worldview to a bunch of state legislators mostly from coal counties. We will present our ideas about renewable energy entirely within the existing paradigm. We will say that solar saves money, and that is the reason to do it. The real costs of fossil fuel energy use—environmental pollution, carbon emissions, and increased health care costs—are not included in the price of coal. If you include them, coal is way more expensive than solar. But we won't say even that. We won't say the "c" word, because they will circle their wagons. We will be polite to coal that day. We will tell them coal is fine; we just want a little tiny part of the action and will let coal have all the rest. They will nod their heads politely, ask a few questions to show they are listening, and then go on to the real business of state legislation.

The Climate

March 20, 2012 – Louisville, KY
High 85°F – Low 65°F – Precipitation 0.0 inches

It's a glorious day. The sun is shining, there's a soft breeze, windows are open, birds are singing, and fruit trees are blooming. The quince bush is bright red-orange, the forsythia is singing in yellow, and redbuds are in their purple glory. The lawn is a dark, early summer green, speckled with happy-face dandelions. It's a little hot, supposed to get up around 86 degrees this afternoon, but not bad for the *last day of winter.*

IT HAS BEEN BEAUTIFUL FOR WEEKS NOW: SEVENTIES AND eighties mostly. It's expected to stay that way for the next ten days, with highs a bit hotter tomorrow. I was driving on the interstate the other night in a T-shirt with the window rolled down. The weeds have been growing in my garden all winter long. I normally put eggplant in the ground in May, but I feel like I could put it out there now. Most of the wildflowers in the woods are already in full bloom, and the daffodils have come and gone. The old-timers in these parts call daffodils *March flowers,* but they have been blooming in February the last few years, and this year they bloomed in January. Their shoots came up out of the ground in December. The daffodils don't need to be convinced of anything. They know what's going on.

Tornadoes are normal in spring and summer, but in the *winter*? There were massive tornadoes here a couple of weeks ago—twenty-two people killed in Kentucky. The arctic ice cap is likely to disappear one summer within my lifetime. What's going on here? What kind of evidence do we need?

A lot more, it turns out. I can point to what is going on outside my window right now, but about this time three years ago, when we were trying to shut down the Capitol Power Plant in Washington, it was freezing cold. We were walking through snowbanks. People laughed at us for trying to get arrested in the name of "global warming." Weather, of course, is not climate. Climate is *average* weather. You determine an average by adding a column of figures and dividing by the number of figures. You can get the average you want by choosing which figures to add. If I find a place with a string of unusually hot days and add up the figures, I'll come up with something frightening, while you can pick another time and place to "prove" exactly the opposite. This year, while we were sweltering all winter long in the lower forty-eight, Alaska had record snowfalls and Europe experienced one of the coldest winters on record. Could the climate be going the other way, perhaps?

No. It's definitely getting warmer. If you add up figures for the whole world over the last few decades, there is no doubt temperatures are rising. The climate is changing, right now. But will it continue changing? We could have a few cold years and bring the average back down again. The short-range trend is clearly hotter, but short-range weather trends are not climate, either. How far back and how far forward do you go to know what is really happening? Even if you get good average figures over a long range, you're still dealing with averages. It could be way-off-the-charts hot on one continent and back-to-the-ice-age cold on another. That would be a trend in itself, but the worldwide average would show up "average."

Climate is like the night sky; we are all looking at the same thing while seeing what we are conditioned to see. I think it's changing. I see it happening all around me, and I like to hear stories in the media confirming my worldview. But I'm vested in that worldview. I live it, I work it, I write books about it, but I don't really know. I might be cherry-picking what I want to know.

And nobody will know for a long time—until well after the current generation is gone. Even if we did know, we could not be entirely sure climate change is caused by human activity in the form of carbon dioxide emissions. We all know CO_2 is a greenhouse gas—we know it heats the atmosphere—and we know there is 40 percent more of it in the atmosphere now than when we started puffing it into the air. But we still can't be sure we are the ones causing the planet to heat up. There are natural forces constantly releasing and absorbing carbon to and from the air: volcanoes, seeping methane, limestone formation, and sea floor subduction, for instance. And even if we knew we were the cause, we would not know if there were anything we could do to reverse it. We could go to a lot of trouble to improve our ways and still find the climate changing all around us.

Besides, what's so bad about warm weather in March? It's beautiful out there today!

LET'S TAKE A CLOSER LOOK. OBVIOUSLY, WE WANT TO KNOW what the climate is going to do in the future, but we can only know for sure what it has done in the past. That's a good place to start. But not last year, or when your grandmother was born. We need to start way back in geological time. The further back we look, the bigger the picture becomes, and the bigger the picture, the more we will know how to fit into it.

About six hundred million years ago, before plants and animals evolved, there is evidence that the climate was extremely cold. Ice caps developed at the poles and steadily grew toward the tropics, to the point where some climatologists believe there was only a narrow band of ice-free ocean along the equator. No one is sure why—perhaps there was not enough carbon in the air—but the cold snap lasted for several million years (like everything else in geological time) before things warmed up again.

Things had warmed up considerably by the time animals evolved 540 million years ago. There was plenty of oxygen dissolved in seawater and nearly as much in the air as there is now. Photosynthesizing plants soon sprouted on the land, keeping the oxygen level high and the carbon level relatively low. The result was a balanced ecosystem and a balanced climate. But the climate fell out of balance from time to time. There was a long period of time (the Carboniferous Era) two hundred million years later when plants got way out ahead of animals. Large animals had not yet evolved to eat plant foliage, and runaway photosynthesis was absorbing carbon from the atmosphere and storing it up in plant leaves and branches. As continents shifted and sea levels rose and fell, seawater flooded swamps and forests and buried them in sediment. Billions of tons of carbon in the plant material were trapped below ground away from the air. So little greenhouse gas was left in the air to trap the sun's heat that the climate turned colder and colder. This is the same carbon we are now digging out of the ground and reintroducing to the atmosphere, making the climate hot again.

This is too simple a picture, of course, but generally accurate. The coal, tar sand, natural gas, and petroleum we now burn were largely laid down hundreds of millions of years ago. Taking carbon out of the air then made the climate cooler; putting it back in now makes the climate warmer. That much is pretty clear.

Two hundred and fifty million years ago, the greatest mass extinction of all time killed almost everything on Earth. Life as a whole nearly came to an end. Scientists are fairly certain the die-off was climate related, and pretty sure the culprit was the Siberian Basalt Traps, a continent-sized volcanic deposit thousands of meters thick covering most of western Asia. Like a volcano without a cone, the entire region erupted with an upwelling of basalt, spewing tons of

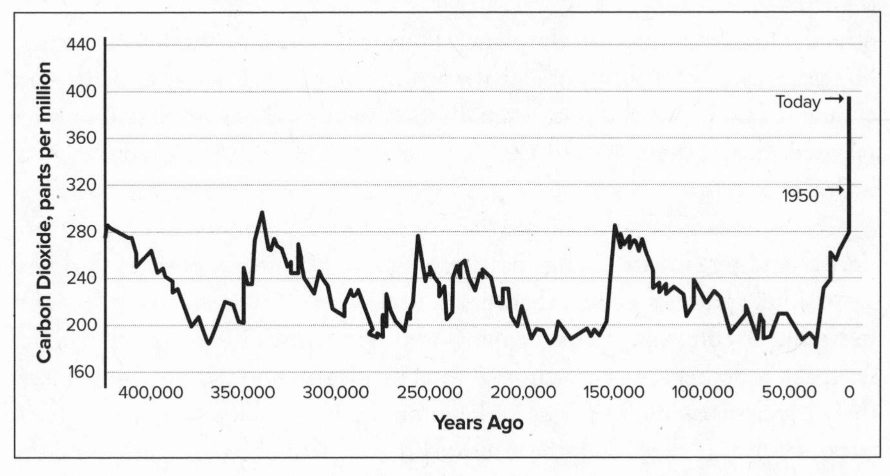

The level of carbon dioxide in the atmosphere is unprecedented. (Source: NOAA)

climate-killing carbon dioxide and sulfur dioxide into the air. Airborne debris darkened the skies, and global temperatures cooled. Then, as debris settled and greenhouse gases set in, temperatures skyrocketed. The whole planet became too hot to handle. Sea levels plunged, then overflowed. Acid rains fell all over the world, plants and animals died everywhere, and continents were stripped bare of topsoil.

Almost everything died. Only a handful of organisms survived to replenish the earth. Somewhere between 90 and 95 percent of all species became extinct on land, in the air, and in the sea. We don't know how fast this happened, but it probably did not happen overnight. More likely, trends developed and continued over a ten-thousand-year period. The problem of understanding the climate then is the opposite of understanding it now: we have averages for climate change over millions of years, but we do not know what is inside the averages. We don't know what happened year-to-year or century-to-century. Now, we know what happens every day, but do not know what it means in long-term averages.

One likely factor in the climate disaster that caused the extinction is a hydrocarbon called *methane hydrate*. It consists of methane (natural gas) locked up in frozen water molecules. It rests in the ocean's depths even now, frozen and undisturbed, in both the arctic and antarctic polar regions. Every hundred million years or so it thaws, wakes up, and burps billions of tons of greenhouse gases out into the atmosphere, toasting the earth's surface. Methane does not stay in the air as long as carbon dioxide, but while there it traps *twenty times* as much heat from the sun. Because there is so much of it and because its effects are so drastic, methane hydrate produces the worst runaway greenhouse effect we know. There is twice as much of it sitting on the ocean floor as there are all other hydrocarbon fuels on the planet. During the great extinction of 250 million years ago, other forms of global warming may have heated the planet just enough to thaw out and release small pockets of methane hydrate. But once released, the methane heated the planet even more, which thawed out more hydrates, and so on.

This is what is known as a *positive feedback loop*, an important concept for understanding climate change past, present, and future. A positive feedback loop, simply put, is an effect that makes itself worse. Once started, it feeds on itself and becomes increasingly dangerous over time. The methane hydrate feedback loop begins as the climate reaches a warm enough tipping point to thaw the first few molecules deep down on the ocean floor. It is *positive* feedback because, as they reach the atmosphere, the methane molecules cause the warming that causes more thawing that causes more warming. The methane hydrate feedback loop is rare because it takes a lot of heat a long time to reach the ocean depths where the methane is frozen, but once started, there is no way to stop it.

According to scientists, there are at least three other positive feedback loops associated with the earth's climate: *ice melt*, *carbon cycle*, and *permafrost*. The first, the ice melt loop, is already in effect. Sea ice helps keep the earth cool because its shiny surface reflects 80 percent of incoming solar energy back into space. As ice melts, the dark surface of open water reflects only 5 percent of the sunlight, absorbing the other 95 percent. Over vast expanses of the earth's surface, this increased absorption heats the planet even more, which causes more ice to melt, which leads to more absorption, etc. Once it gets started, it is hard to stop, as we are now finding out.

The second, the carbon cycle feedback loop, begins with the large-scale destruction of plant life and the general warming of soils all over the earth. Cutting down tropical rain forests, for instance, releases huge quantities of carbon into the air from decomposing plant materials, while at the same time reducing the rate of global photosynthesis that removes carbon from the air. This further heats the planet, which in turn destroys more biomass. Even without logging, the carbon cycle feedback loop can be triggered by strengthened and protracted El Niño events caused by waters warming in the eastern Pacific. The Amazon rain forest depends on easterly trade winds bringing in moisture from the Atlantic Ocean; El Niños warm eastern Pacific waters and reduce the easterly winds, bringing less rain to the Amazon basin. As trees die, they release carbon stored in their trunks, leaves, and branches, which becomes more greenhouse gas, and in turn further heats the general climate. Warmer temperatures cause a relatively small tree loss initially, but because each tree holds moisture for neighboring trees, its loss can begin a chain reaction that will dry the entire forest.[9]

There is also a vast amount of carbon trapped in organic matter in the soil—probably twice as much as is in the atmosphere now.[10] As soil warms, bacterial activity increases, which releases more carbon into the air and causes more warming. The carbon cycle loop in each of its forms is likely to kick in by the mid-twenty-first century with continued tropical deforestation and global temperatures rising by about 3 degrees Celsius. Once it begins, it may release enough carbon to bring the atmospheric level up by as much as an extra 250 ppm, which would bring global temperatures up a total of about 4 degrees.[11]

The third feedback loop, permafrost, is similar to the methane hydrate loop in that it involves the release of frozen hydrocarbons (methane and carbon dioxide) into the atmosphere, but it refers to thawing soils, not sea depths. The permafrost loop comes into effect much more quickly than methane hydrate (the fourth feedback loop) because soils warm more quickly than ocean depths. There is some evidence that the process has already begun, but it is expected to become a major factor only when global temperatures rise into the four-degree range.

These feedback loops are what we most want to avoid. Once the global temperatures reach a certain level, they may kick in the first loop, which may warm the planet enough to trigger the next loop without our doing anything else. This is "runaway" global warming. Based on threshold temperatures, the likely sequence is: ice melt (now in effect), carbon cycle (an increase of three degrees), permafrost (four degrees), and methane hydrate (five or six degrees for an extended period). The Canadian tar sands could release enough carbon dioxide into the atmosphere to initiate the second feedback loop, and there is no telling what would happen from that point on.

There are also *negative* feedback loops. Rising carbon dioxide levels, for instance, cause more plant growth, which extracts carbon from the air, tending to restore a cooler balance. Carbon in the shells of calcareous animals (coral and shellfish) is also sequestered from the atmosphere in limestone formations. Both of these can help keep carbon levels low, but both are now diminishing rapidly with deforestation and acidification of the oceans.

Three million years ago, just before the ice ages, a period of global warming began that scientists find interesting because of its similarity to what we may be moving into now. Carbon levels were only 360 to 400 ppm—just about where they are now—but the average temperatures were a full three degrees warmer. This might not sound like much, but it was enough, over time, to bring tropical forests as far north as Canada and allow plant and animal life near both the North and South Poles. This was the last time the North Pole was free of permanent ice. The question is, how did the climate become so radically different then from what we experience now with only a little more carbon in the atmosphere? The answer is probably *thermal inertia*, another important concept in climate science.

It takes decades, centuries, or millennia for changes in atmospheric chemistry to show up in the weather. It takes time for things to heat up. Air warms more quickly than water because the atmosphere heats from the bottom up—from the earth's surface up through the troposphere. With atmospheric warming, there is a time lag of decades: the carbon we release into the air now, for instance, may not show up as warming until twenty or thirty years from now. The warming we are seeing now is probably from carbon our parents and grandparents released many years ago.

The oceans take longer to warm—centuries and millennia—because they are heated from the top down, like trying to boil water with a fire above the pot. Water also holds more heat than air and takes longer to heat in any case. Three million years ago, the atmosphere was only three degrees hotter than now, but it stayed warm for a long enough period of time to warm the oceans and melt the poles. This changed the climate everywhere and brought vegetation to polar regions. Climate change was a result not only of temperature changes, but of *sustained* temperature changes. (Once the oceans became warm, it took an equally long period of time for them to cool down again as the ice ages began.)

Another reason the climate was so drastically different then with only a slight temperature difference is the carbon cycle feedback loop. This loop was probably in effect in the tropics three million years ago, and it may be the direction the climate will go again if we let it get three degrees or more above the current level. One way this could happen would be for increased temperatures at the surface of the oceans to create a permanent El Niño effect over the Pacific. That would eventually dry up the Amazon basin and lead to collapse of the world's largest rain forest. With emissions continuing as they are, we could reach that three-degree rise by 2050.

The ice ages (Pleistocene) began 2.5 million years ago, about the time *Homo habilus* was shaping the first stone tools in eastern Africa. Ice caps formed at the poles and have remained there ever since. There are many possible explanations for this strange icing over of the planet so recently in geological time, but none of them seems to be atmospheric. Scientists think the cause is most likely astronomic—a change in solar radiation or a change in the earth's orbit. Or it may be changes in ocean currents. No one knows for sure. Ice ages have been coming and going in fairly regular patterns, and some variation of the pattern will likely continue into the future. Each cycle lasts about 150,000 years. We have been in an interglacial period for the last 10,000 years and are due for another ice age a thousand years or so from now.

So, what will it be, global warming or ice age?

Global warming, for now. The daffodils and ice caps are shouting that in our faces. Global cooling may offset global warming at some point in the next few thousand years. The earth may warm up or cool down several times over the next ten thousand or million years, as it has so many times before. But what we have to face now is global warming. We have to think about the next one hundred years.

IF GLOBAL WARMING HAS HAPPENED SO MANY TIMES BEFORE, what's the big deal now?

The big deal now is the *rate* of warming. Previous climate changes have happened over thousands or millions of years (at "glacial" rates). This one is happening in decades—much too fast for life to adapt. Carbon levels were much higher during global warming events of the past, but the rate of increase in carbon levels is ten to one hundred times faster now than *anything we know of in the past.* The climate is probably changing faster today than at any time in the earth's history.

And the rate of change is itself increasing. It used to be thought that arctic sea ice would disappear in the summer some time around the end of the century. It is melting so fast now that it may be gone in twenty to thirty years. Birds don't know when to migrate; insects don't know when to hatch. Trees don't know when to leaf out. Five years ago, it was so warm in early April where I live that all the trees went into full foliage. Two weeks later, the temperature plunged into the mid twenties Fahrenheit four nights in a row, and all the leaves in the forest turned black. Climate change is so rapid now that animal species do not have time to migrate northward or uphill. Seed dispersal for most forest trees allows forests to migrate at about one mile per year, much too slowly to outrun our current rate of climate change. Plus, habitat is so damaged and fragmented that there is nowhere to move to and no way to get there. Tropical forests used to spread slowly northward as the climate warmed over thousands of years. Now they are disappearing before they have a chance to go anywhere.

The *really big,* big deal is agriculture. The human species has survived all kinds of climate chaos over the last two and a half million years, but human *civilization* has not. Agriculture, the ecological basis of human civilization, is less than ten thousand years old. It has existed only since the end of the last ice age. The climate throughout this period has been relatively stable. Agriculture has known nothing else. Starting out in a small way on the river plains of Mesopotamia, the Nile, the Indus, and the Yangtze, agriculture has since spread to cover nearly all of the earth's surface to which it is suited. Each local agricultural mix is fine-tuned to its altitude and latitude, and each is carefully adjusted to its local climate. Agriculture cannot migrate to more favorable locations because, with few exceptions, it already exists at all locations.

We are beginning to see more heat; more powerful hurricanes, tornadoes, and hailstorms; longer, hotter droughts; bigger floods; even a "superstorm." This is not good for a stable food supply. The rate of climate change is so rapid that human food supplies cannot be sustained for much longer, even if the world population were to remain stable. Deserts are spreading into croplands in Africa, Australia, Asia, and North America. Mountain snowpacks that provide summertime irrigation waters for many parts of the world are melting rapidly. Storm surges and rapidly rising sea levels are bringing brackish seawaters into cultivated river deltas. For every degree over 30 degrees Celsius, rice, wheat, and corn production declines by 10 percent. Over 40 degrees, there is no production at all.[12] Food prices are rising sharply for those who can pay them. For many people, paying more is not an option.

If we were still hunters and gatherers, we would move with the herds, learn of new edible roots and berries, and adjust however necessary to this climate change as we adjusted to climate changes in the past. But as urbanized, agriculturally dependent societies, we can no longer migrate with the herds or find new food sources in the next river valley. A stable food supply depends on stable agriculture, which depends on stable climate. Changes in temperature and rainfall along with extreme weather events will reduce world agricultural output. As changes become more severe, food shortages will become more severe. Agricultural production is barely able to keep up with population growth even without climate change; as climate change progresses, food supplies will decrease while population increases. Large-scale famines will become all but inevitable.

IT IS WARMER NOW THAN AT ANY TIME IN MORE THAN A thousand years. Arctic temperatures are up 2.3 degrees in the last fifty years. The northern ice cap is losing 100,000 square kilometers of ice per year, initiating the ice melt feedback loop. In the last thirty years, both jet streams, the weather drivers of the southern and northern hemispheres, have shifted one-degree latitudinally toward the poles. The subtropical heat zone of northern Africa is moving up to the Mediterranean and southern Europe. Seventy percent of the coral reefs on Earth are dead or dying. Sea levels are rising at a rate

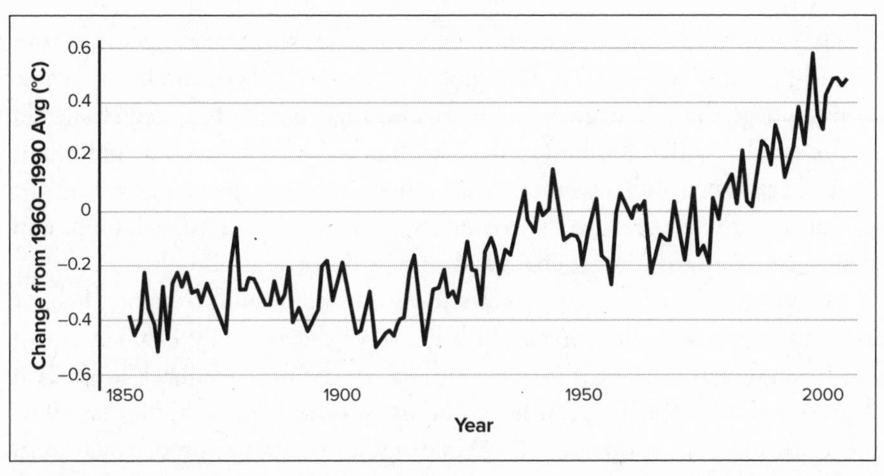

Global average temperatures have been rising for a hundred years. (Source: NOAA)

of 3.3 millimeters per year, faster than the 2001 Intergovernmental Panel on Climate Change (IPCC) estimate of 2.2 mm per year.[13]

To understand what is likely in the coming years, climate scientists enter data on carbon levels, temperatures, storms, and sea levels into complex software programs that show how these factors are likely to interact. Results always depend on inputs, and wide ranges of uncertainty remain, but there is an emerging picture. No one can say exactly how high average temperatures will go, but we have some idea what will happen to the world at various temperature ranges and carbon levels. We can say, for instance, that if carbon emissions continue to increase, rising sea levels will put Miami, New York, London, Bangkok, Shanghai, and Mumbai underwater and send fully one half of humanity scurrying for higher ground by 2150. But these rises in sea levels are fairly long range because of thermal inertia. More likely to be seen first are the atmospheric effects of climate change, such as hotter average temperatures, droughts, flooding, megastorms, habitat loss, and extinctions. The snowpack in the Himalayas, for instance, where the Ganges, Indus, Brahmaputra, Mekong, Yangtze, and Yellow rivers all originate, is already melting. These rivers together supply water for half the world's population. Bangladesh will be hit four times over: river flooding from melting snows in the Himalayas, stronger monsoon rains, higher storm surges, and generally rising sea levels.

The overall picture looks something like this: in the past 150 years, since fossil fuels came into use, atmospheric carbon levels have risen from 280 parts per million to 393 ppm. The average global temperature in this time span has risen 0.8 degrees Celsius. There is another 0.5 to 1.0 degree increase on its way due to thermal inertia and to the ice melt feedback loop that is already in effect. This much warming will happen even if we stop all carbon emissions tomorrow. The IPCC estimates that by 2100, global temperatures will rise by anywhere from 1.4 degrees to 5.8 degrees—a very wide range. A rise of 1.4 degrees would be noticeable, while 5.8 would be disaster. But here's the take-home point of it all: if we can keep global temperatures from rising more than *2 degrees Celsius* (3.6 degrees Fahrenheit), we will avoid the worst positive feedback loops and avoid a runaway greenhouse effect and climate disaster.

An increase of two degrees will still mean storms and droughts, habitat loss and extinctions, flooding and likely loss of the arctic ice cap, but it will not mean global disaster if it does not kick in the next feedback loop. Two degrees will wipe out the polar bears and many other species, but it may allow most other species to survive. The Maldives, an island nation of 269,000 in the Indian Ocean with a highest elevation of only twenty feet above sea level, would like to see the target moved to 1.5 degrees. One and a half degrees might save the Maldives, for a while, but that target seems way beyond anything politically achievable in a world that does not generally recognize what is going on. Two degrees is only slightly more realistic, and only slightly below critical threshold levels, but it may be the best we can hope for.

Many say the two-degree target is unattainable in the real world of unbridled fuel consumption and continuous economic growth. They may be right. But two degrees seems to be the built-in limit, when you do the math. It's a target proposed by climate scientists and endorsed by the European Union, the Group of Eight, and many environmental and humanitarian groups. It's not a perfect target, and there will be a lot of damage even if we achieve it, but we can live with it ... most of us.

The German Advisory Council on Global Warming suggests that we look at surviving climate change in terms of a long-range carbon budget. For a two-in-three chance of keeping warming to within two degrees, humanity's total worldwide carbon budget through the year 2050 should be 750 gigatons of CO_2. The estimated world population for that year is 6.9 billion people. To stay within budget, carbon emissions have to be no more than 2.8 tons per person per year.[14] There are people in developing countries already using less than this, but the average American uses about twenty tons per year, or seven times too much. Humanity has already spent 30 percent of its twenty-first century budget in the first ten years.

So how do we keep climate change to two degrees? The computer models say that if we keep carbon levels below 400 ppm we have a 75 percent chance of preventing global temperatures from rising more than two degrees. But levels are already at 393 ppm and rising 2 ppm every year. This means that emissions *will have to peak by 2015* and begin to decline thereafter. That may be impossible to accomplish. Some economists say that anything under 550 ppm is not "economically realistic." They may be right, but at that level of carbon, we have only a 20 percent chance of keeping below the two-degree increase and avoiding the more serious feedback loops. The earth's climate system, it turns out, is built on physical law, not "economic reality."

An increase of more than two degrees means we risk a runaway climate and global foreclosure in the not too distant future. The carbon cycle feedback loop will kick in, and the Amazon will become a desert. There is evidence that this is already happening. A massive drought hit the western Amazon in 2005, killing billions of trees and dropping river levels as much as forty feet. It was called a once-in-a-hundred-years event, but five years later, an even more severe drought shriveled billions more trees, releasing unprecedented tons of stored carbon into the atmosphere. The Amazon, the world's greatest living carbon sink, became for the first time a net *producer* of atmospheric greenhouse gases.[15] Continued release of so much fixed carbon and loss of so much photosynthesis may lead directly to a four-degree temperature rise, which would kick in the next positive feedback loop: the *permafrost.* Enormous deposits of carbon dioxide and methane frozen in the arctic tundra soils of Siberia, Alaska, and northern Canada would begin to seep into the atmosphere—one and a half trillion tons of them. This is the equivalent of twice the current amount of carbon dioxide in the world's atmosphere. As the climate warms, more permafrost melts, and more gases are released into the air. This traps more solar energy and further warms the climate, which melts more permafrost, and releases more gases, and there you have the feedback loop. The more warming you have, the more warming you get.

The permafrost loop could bring worldwide temperature increases into the five- or six-degree range, which, over an extended period, could kick in the methane hydrate loop—our ticket to mass extinction.

CLIMATE DISASTERS AND THE FALL OF CIVILIZATIONS ARE A normal part of history. Species come and go, empires rise and fall, and life goes on. But there is something very different about this particular looming disaster: we are not bystanders. Climate change is not happening *to* us. We are players in the game, not innocent victims. We are creators of the disturbance and *aware* of the disturbance. We see it. For the first time in the history of the living world, consciousness is playing a role in a biospheric crisis. You, reading these words, are part of it.

To live successfully with other life, we will have to do more than reverse the damage that has been done. We cannot simply go back to our preindustrial selves; we are too big for that. Our economic interdependence is too complex to unravel, and there is no practical way to deindustrialize. We will not go back. Life does not go backward.

Human awareness is expanding rapidly into the inner workings of the terrestrial system; we are gaining confidence in being here and learning to act for the good of all. Other forces of nature act as well: volcanoes erupt, forests photosynthesize, methane seeps from tundra soils, and ocean currents stream northward. No forces of nature *control* the weather; but each affects it. Humanity will affect the climate from now on and must learn to embrace that reality. Climate policy will change as the world changes: there may be times when we deliberately pump *more* carbon into the air to warm the place up. But we must learn to do it properly. We will know how what we do changes weather patterns, caribou migrations, bacterial activity in the soil, agriculture, and human life expectancy. We will learn to balance looking out for ourselves with looking out for the needs of other living beings.

We need a detailed, carefully constructed, worldwide climate policy. We need a direction, a target to aim at—something like 2 degrees Celsius and 350 ppm of atmospheric carbon dioxide—something we can live with. But more immediately, we need a worldview within which to have the policy. If there is a tiger lurking in the forest nearby, we need to decide what to do, and do it quickly.

The climate change deniers don't want us to move. Some of them honestly believe there is nothing to fear, and I respect that. But many are subject to manipulation by enormous financial interests. Merchants of carbon-based fuels are *the largest corporations in the world*. Their vision is so clouded by money that they seize and magnify any shred of evidence that casts doubt on climate change—and there is plenty of it. Not evidence that climate change is not happening, *doubt* that climate change *is* happening, or doubt that it is caused by fossil fuel combustion. Doubt is a powerful weapon in their arsenal because it prevents decisive action. Doubt freezes us in the eyeballs of the tiger looming toward us. The deniers no longer deny that the tiger is there, they just don't want us to look. They want us to stand there, in the middle of the forest, waiting for more evidence that he will leap.

Back in the nineties, the deniers formed something called the Global Climate Commission, a group of fossil fuel advocates and a few rogue climate scientists funded by ExxonMobil and other carbon-based energy companies. Their goal was, and is, less to refute climate change than to produce enough counterevidence to keep the public and elected officials from doing anything decisive. They want to keep the idea alive that there is "debate" among scientists and that we should do nothing until all the facts are in; that is, after it is too late. Internal documents reveal that as far back as 1995, their own scientific advisers informed the leadership that man-made global warming was "well established and cannot be denied."[16] The board of directors removed this bit of embarrassment from the commission's public statements, and the commission has continued polluting public opinion ever since. There is not much to be said about these people now. They will continue to squawk as long as anyone listens, but they become more ridiculous as time passes. They are the Flat Earth Society of the economic paradigm.

The earth is not flat; the climate is changing; and the change is a result of combustion of carbon-based fuels.

March 22, 2012 – Frankfort, KY
High 79°F – Low 57°F – Precipitation Trace

Back in Frankfort this morning for the committee hearing. Fifty or so of us packed the room in support of the Clean Energy Opportunity Act. A handful of people from utilities and coal companies were there to oppose it. The committee chair gave us a half hour to explain the benefits of renewable energy portfolio standards, energy conservation, the feed-in tariff, expanded net metering, the health benefits of solar and wind power over burning coal, the environmental benefits, the climate benefits, the employment benefits, and the long-range utility rate benefits. The committee listened, and we got about halfway through the beginning of our presentation before time ran out. One committee member commented on how cheap electric rates were in Kentucky compared to states with renewable energy legislation. The chairman noted that all forms of energy are good and have their place, and he concluded with the remark that, "This has been an interesting discussion." No vote was taken, and the committee returned to its regular business.

President Obama is in Oklahoma today and will sign an executive order expediting construction of the southern leg of the Keystone XL pipeline from Cushing, Oklahoma, to Port Arthur, Texas. It's easy to see where this is going. How can he approve one part of a pipeline and not the other? There will be some rerouting through Nebraska, but we're going to have that pipeline.

The Ninety-Nine Percent

March 26, 2012 – Louisville, KY
High 68°F – Low 48°F – Precipitation 0.0 inches

The dogwoods are already in bloom and will be gone in two weeks. The Dogwood Festival in Louisville is scheduled for April 20, three and a half weeks from now.

I WAS IN ST. LOUIS OVER THE WEEKEND, TRAINING TO BE A NVDA (nonviolent direct action) trainer. The two-day session was sponsored by the 99% Spring, a group of seventy-five local, national, and global community action, union, and environmental organizations, including MoveOn.org, the Rainforest Action Network, the United Auto Workers, 350.org, the AFL-CIO, Jobs with Justice, Greenpeace, and the United Steel Workers. The group was formed less than six weeks ago for the purpose of training one hundred thousand people around the nation to take part in direct actions this spring. I and four others from Louisville were to be trained to train others when we returned home.

We met at the Teamsters Local 688 Union Hall. There were forty-some people in the room: black, white, Latino, and Asian, aged twenty to seventy, women in pants to pantsuits, men with buzz cuts to ponytails, dressed anywhere from T-shirts to button-downs, and not a tie in the room. No one-percenters here. UAW men wore T-shirts reading "Got Union?" and "Management Is Not Your Friend." One woman sported a large, frizzy Afro. Professions ranged from students, printers, teachers, and occupiers to community organizers, autoworkers, electricians, and retired postal workers. Anyone without a large stock portfolio would have felt at home. Everyone was pleasant, smiling, eager, enthusiastic, energized, tolerant, and ready to reach out to new friends and allies. This was something new.

The concept of the one percent is an excellent organizing tool. It focuses attention. It brings everyone else together. The rest of us, whoever we are, have in common *being the ninety-nine percent*. That's enough of the total to push this country in any direction we want. Hey, if we want guaranteed employment, guaranteed housing, universal education, universal medical care, clean water, clean air, a stable climate, and safe neighborhoods, we can make it happen. We have the resources to do it, and working together, we can get it. The people united will never be defeated! We are the ninety-nine percent! Nothing can stop us!

But the one-percent guy is going to win in the end because, should you ever defeat him, you become him. You have to fill the vacuum he leaves by losing. If you defeat the ruling class, you become the ruling class. If you overthrow the government, you become the government. If you topple Saddam, you become Saddam. If we're talking about a reform movement here—if we're throwing out the bathwater and keeping the baby—we're not really trying to convince ourselves of how bad the one-percenters are; we're trying to change the rules that give them such an advantage. We're getting smart about things such as deregulation, campaign finance, taxes, government bailouts, and corporate welfare. We can demonize the one-percenters, but we need people running large organizations, hiring and firing, moving capital, and evaluating credit risks. We don't need deregulated banks taking private gain with public money, but we need people who can figure out what is a good investment and what is a bad one. Private enterprise or public enterprise, private capital or public capital, we need people specializing in credit analysis, investment strategy, and financial transaction.

The one percent makes itself vulnerable to revolutionary pressure by mistaking class interest for economic freedom. It often sees itself as magnanimous—heroic, even—providing for the rest of society by exercising its own economic freedom. Money is made in the course of helping people get what they want. Wealth is a reflection of how much one has contributed to society. Therefore, it comes to believe that its enormously privileged position is a benefit to everyone, not just itself. "The lower *our* taxes, the more money we have for investment, and the more jobs we create. When we're better off, everyone is better off." This leads to the absurdity of millionaires and billionaires with multiple megamansions paying lower tax rates than the people cleaning their kitchens and mowing their lawns. What kind of society would do that? How did we ever decide to make capital gains taxes *lower* than wage taxes? What made us think that the guy digging ditches should pay a *higher* rate than the guy who makes his money by having money? What kind of justice is that?

If the one percent continues to press its advantage through unfair taxes, superPACs, bailouts, and corporate welfare, there will be class warfare and revolution. Justice demands it.

The downside of singling out the one percent is that it divides humanity in two. The ninety-nine percent is most of us, but it is not all of us. It does not include leaders in government and business, and many others making decisions that direct the overall course of human evolution. Setting the one percent outside the pale is no better than excluding any other class, nationality, gender, race, preference, or profession, and possibly worse.

Division blocks vision; *other people,* no matter how few, become the focus of action rather than people as a whole, and the larger picture is lost. People driving cars will continue to blame oil companies for climate change, and people with computers and hair dryers will blame coal companies for mountaintop removal. Singling out the one percent pits us against each other and keeps our minds within the economic paradigm. Until we see ourselves as a whole, there will be no way to fit the economy within the ecology.

SO, THERE I WAS, SITTING IN A ROOM FULL OF NINETY-NINE-percenters, wondering how I ended up there. I want to stop a pipeline, not overhaul the banking system. And I began worrying about how these people were going to see me and my type. Fussing about clean water and atmospheric carbon can seem distant when you're dealing with hunger and bankruptcy. And what were those union guys over in the far corner talking about? They were going to be interested in higher wages and more jobs. What interest were they going to have in *not* building something? I decided to ask.

Mike looked like an autoworker, the kind of guy with little patience for peaceniks and tree huggers. But he was proud of the United Auto Workers' environmental and civil rights record. "We were the first big union to accept blacks," he said proudly. Walter Reuther was his hero. He told me how Martin Luther King often used the UAW's Solidarity House in Detroit as a refuge when things got crazy in the South. "He wrote the *I Have a Dream* speech there, right there, as a guest of the UAW. We even posted bond for him when he got arrested," he said.

Mike was motivated by a strong sense of justice, not just bread and butter. He was chairman of UAW local 2250's Community Action Program, an outreach program that deals in non-job-related issues. He was particularly interested in voter registration and getting people out to the polls. His activism often goes beyond union membership. "I can't always speak in the name of the UAW, " he told me. He was passionate about getting jobs for people, even nonunion jobs. "If somebody has a job at Walmart, that's good. I might feel sorry for him, 'cause it's a low paying job, and he would be better off with a union, but having a job takes care of a lot of problems, a whole lot of problems, and Walmart is the biggest job provider in the world," he said.

He was against the Keystone XL pipeline for environmental reasons. "There could be spills anywhere," he said. "We need jobs, but not those jobs. The pipeline would not be good for the country. There are a lot of unions in favor of it, and I don't blame them, but we don't need jobs doing the wrong things." He knew the tar sands petroleum would be on the world market and not exclusively reserved for U.S. domestic consumption, but he did not know about the environmental disaster of tar sands extraction or about the carbon bomb. As with most who are aware of the pipeline, Mike had heard nothing about the wider environmental consequences of extracting and burning so much fossil fuel. "It sounds like another way to keep the price of gas down so we won't develop other forms of energy," he said.

He was torn about the problem of producing less waste while maintaining employment. "GM makes a six-year car. ... It ought to last longer than that, but we have to keep people employed. I don't know what to do about that," he said. But Mike was enthusiastic about the ninety-nine percent: "We all have to do this together. If we can get this coalition together, we can make something happen in this country."

Talal was equally enthusiastic about the ninety-nine percent. He came from North St. Louis, which he said was "the poorest community in St. Louis." But Talal got a big break when his mother was able to get him into school in the suburbs. "I stayed overnight in peoples' houses near the school," he told me. "My eyes were opened; my horizons were broadened. I experienced some racism but saw a new world of possibilities. Most people were really nice and treated me well. But I lived in two worlds. I was underprivileged at school and overprivileged when I went home. Even the languages I spoke were different. I learned to articulate my thinking in school, but my friends thought the way I talked sounded odd."

Talal, like Mike, was motivated by justice issues, but he had a side interest in the environment. "The ninety-nine percent is potentially much more powerful than the one percent," he said. "It's like my dog. He's really strong, but I can control him with a little string around his neck. He could easily break the string, but he won't." When I asked Talal if the pipeline issue would be part of the 99% Spring, he was doubtful. "The ninety-nine percent has a specific agenda. I don't know if the XL pipeline is in the agenda," he said. Talal is now a fine arts student at St. Louis Community College, Forest Park, and thinking about law school.

THE NINETY-NINE PERCENT COALITION IS TOO WIDE AND NOT quite wide enough at the same time. It is something that has never happened before, and it may do things far beyond anything imaginable now. The coalition pools so many conflicting causes that it is bound to fracture at some point, for instance, between unions and environmentalists. Both are eager to work out differences and pull together against the corporate bad guys, but what happens when there are a thousand good union jobs staring at a pipeline project? I'm in construction. I know how frustrating it is to have a project all lined up and ready to go—big money dangling in front of your nose—and some little thing comes along and blows the whole thing up, some little concern like a few degrees of temperature fifty years down the road. That job is *now*, and I have a mortgage to pay! It is wonderful to see so many people from so many professions, interests, and backgrounds reaching out to each other, wanting to understand each other's point of view, and I will do my part to keep the coalition going as long as it will, but it may be too broad a front. I will believe in it as long as I can.

There is something else about the ninety-nine-percenters that came to the surface at this training session: spirit. I don't mean enthusiasm—it had that, too—I mean spirit, as in *spiritual*. This was a spiritual experience. The *culture* of the ninety-nine percent, and of the Occupy movement it springs from, is accepting, tolerant, inclusive, loving, compassionate, and consummately spiritual in the religious sense—in the sense, perhaps, of an early religious movement rather than its later institutional form. People do things for each other. They listen. They try to find meaning. They try to be fair to others they may not understand. Time and space are always available for people to speak who may not be aggressive about speaking in public. The atmosphere is not metaphysically spiritual but interpersonally spiritual. Where other social and political movements have been overly serious, analytical, and ideological, this one is full of humor, tolerance, fun, and not a little bit of pure silliness. No heavy party line here. This is unique. If it lasts, this could become a whole new way to be who we are.

FIFTY YEARS FROM NOW, THE ECONOMY WILL NOT BE AS IT IS now. The earth will not stand it. It will evolve, one way or another, to fit within the ecology. But it will not evolve without economic justice. The ninety-nine percent will see to that.

March 22, 2012 – Louisville, KY
High 81°F – Low 60°F – Precipitation 0.13 inches

The Senate began debate today on a bill that would repeal $24 billion in tax subsidies to oil companies over the next ten years. Unless this bill is passed, the average American family of four will be donating tax money to the largest corporations in the world—about $400 total per family to ExxonMobil, BP, Shell, Conoco, and Chevron, companies that made $137 billion in profits last year. The money will provide "incentives" for these companies to stay in business. The proposed bill would repeal the tax subsidies and allocate the money to develop alternative energy sources and to pay down the deficit.

According to the *Chicago Tribune*, Senate Republicans are allowing the debate "as a way to showcase the Keystone XL pipeline and other GOP proposals." Both of my senators, staunch fiscal conservatives, got in on the showcase. Mitch McConnell said the proposal to repeal subsidies "will lead to even higher prices at the pump." (Apparently, if we give enough tax money to oil companies, they will give some of it back at the pump.) He then added that repealing tax subsidies would "raise taxes on American energy manufacturers instead of moving forward with the construction of the job-creating Keystone XL pipeline." Rand Paul, my other senator, said, "Instead of punishing them [the oil companies], you should want to encourage them. I would think you would want to say to the oil companies, 'What obstacles are there to you making more money?' ... We as a society need to glorify those who make a profit."

One hundred percent of my senators are doing everything they can to make the case for the ninety-nine percent.

April 2, 2012 – Louisville, KY
High 72°F – Low 61°F – Precipitation Trace

"Acknowledging that the atmosphere is warming doesn't make you a liberal," according to Paul Douglas, Minnesota meteorologist and author. In an article in MPR News (Minnesota Public Radio) entitled "A Republican Meteorologist Looks at Climate Change," he calls this spring's weather "off-the-scale, freakishly warm ... The scope, intensity and duration of this early heat wave are historic and unprecedented. And yes, climate change is probably a contributing factor. 129,404 weather records in one year, nationwide? The climate is warming. The weather is morphing. It's not your grandfather's weather anymore. The trends are undeniable. If you don't want to believe thousands of climate scientists—at least believe your own eyes: winters are warmer & shorter, summers more humid, more extreme weather events, with more frequent and intense rains."

The Global Paradigm

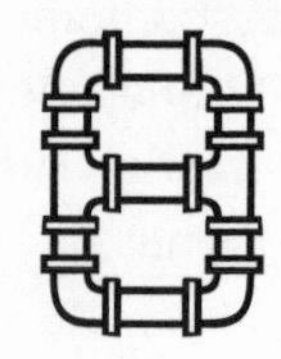

April 11, 2012 – Louisville, KY
High 56°F – Low 35°F – Precipitation 0.0 inches

An Associated Press article by Seth Borenstein reported that, according to calculations by the National Oceanic and Atmospheric Administration, temperatures in the lower 48 states were 8.6 degrees Fahrenheit (4.8 degrees Celsius) above normal for March and 6 degrees Fahrenheit (3.3 degrees Celsius) higher than average for the first three months of the year. That far exceeds the old records. "It's been ongoing for several months," said Jake Crouch, a climate scientist at NOAA's National Climatic Data Center in Asheville, N.C., in the article. Meteorologists say an unusual confluence of several weather patterns, including La Niña, was the direct cause of the warm start to 2012. While individual events cannot be blamed on global warming, Crouch said that extremes such as this are expected to get more frequent because of man-made climate change. It is important to note that this unusual winter heat is mostly a North America phenomenon. Much of the rest of the Northern Hemisphere has been cold, according to NOAA meteorologist Martin Hoerling.

April 13, 2012 – Louisville, KY
High 67°F – Low 40°F – Precipitation Trace

After a series of early spring wildfires broke out in twenty eastern states, U.S. Energy Secretary Steven Chu noted in a speech that scientific evidence of climate change is getting more and more powerful: "Over the last couple of years, the dispassionate, hard science evidence has been mounting, increasing." The warm, dry weather across the country, paired with recent wind, "is a perfect recipe for fire" in the eastern United States, Weather Channel meteorologist Mike Seidel told NBC News, as large fires broke out up and down the East Coast this week.[17]

DEEPLY EMBEDDED IN THE HUMAN MIND IS ANOTHER NIGHT sky and another paradigm. The night sky is the world of people; the paradigm is *our* people—people like us, with whom we act, and with whom we are secure. There are as many ways of arranging people into groups—*us* and *them*—as there are ways to arrange stars into constellations. How we arrange ourselves—how we put people into constellations—depends on who we think we are.

And we are political animals. We gather in groups to secure resources and protection. But what sort of groups? Aristotle's "political animal" is more accurately translated from the Greek as "animal who lives in the *polis*." The polis was the city-state, the Greek political unit. What distinguished humans from animals, according to Aristotle, was organized community life within sovereign, independent urban centers such as Athens, Sparta, Corinth, and Thebes. Each city was its own country. This was civilization itself to the Greeks. Our words *political, policy, police*, and *polite* all derive from the ancient Greek concept of the *polis*.

The polis was more than a concept; it was a paradigm, an assumption so deeply ingrained in their understanding of humanity that they thought of it as humanity itself. There was no other conceivable way to construct a civilization. It was how they were defined as a people, but also how they failed as a people. Their first loyalty was to the city, not the nation. This kept them permanently divided, city against city. Exhausting fratricidal warfare between cities became a constant feature of Greek life. Foreign powers intervened on one side or another, exploited civil divisions, and finally dominated the entire country. The Greeks recognized themselves as a nation with a common heritage, a common language, and a common geography, but they never thought of themselves in common *politically*. Cities and regions formed temporary alliances, but the country as a whole resisted any form of national government. The Greeks arranged themselves into beautiful constellations that still shine today, but they never saw the larger pattern that could have saved them from self-destruction.

The tragic flaw in Greek civilization is obvious to us today. We are so imbued with our own *national* paradigm that a short glance reveals to us exactly where they went wrong. Why could they not see it? Why did they not understand that they were destroying their own cities, crops, economies, and creativity with year upon year of senseless interurban killing? They thought of themselves as Spartans, Argives, Megarans, ot Athenians, though plainly, they were all Greeks! The city-state paradigm limited what they saw, what they were able to do, and how long they could survive without foreign domination. Our paradigm would have worked much better for them. But because paradigms are largely unconscious, the Greeks were unaware that the constellations they had arranged were not working for them. They could not see that the city-states that defined their civilization were killing them, because their political paradigm allowed no other possibilities. The vision of a unified Greek nation was dangling right in front of their noses, but they could not see it.

How aware are we now of *our* political paradigm? Are we able to see around it? How well does it serve us, and what other possibilities might there be? These questions are not normally asked, because a paradigm is not a rational decision, not a preference among a set of options. It is something we assume, as a society, and do not generally question. It is something we *are.* If we consider other possibilities—if we arrange the stars into new constellations—we will no longer be *us*; we will be something else. It is this fear of thinking oneself out of existence that kept the Greeks from organizing into a nation-state, and that now keeps us from organizing into a global society. Thinking people today, people who try to view the world from a rational standpoint, generally agree that the world would be better off united than divided into separate nation-states. We would be in a much better position to deal with climate change, resource allocation, economic development, trade regulation, famine, disaster relief, pollution, labor legislation, and the threat of nuclear war from a global rather than a national perspective. But global consciousness does not seem to be something we can *choose.* Even if everyone on the planet suddenly agreed that we should shift political organization from the national to the global level, we would not yet have it. The transition would require more, much more, than rational selection.

But, for the moment, let us see how far reason can take us toward understanding the national paradigm. Let us look at where the national paradigm is, how it works in our world today, and where it may lead us. Then, let us consider the global paradigm, and how it might work. Finally, let us look at how a shift from the national to the global paradigm might relate to a corresponding shift from the economic to the ecologic paradigm discussed earlier.

A PARADIGM, TO BE A PARADIGM, MUST BE ALL-ENCOMPASSING. From within, a paradigm appears to be the world itself, with nothing before, after, or beyond it. If there were a *beginning,* if it had not always existed, there would have to be other possibilities; the paradigm would not be complete.

The national paradigm seems today to be the only possible way to organize civilization. We think of it as *human nature.* But the national paradigm did, in fact, have a beginning, and it was not so long ago. It did not exist for the Greeks two thousand years ago, or for the Romans, or for the feudal hierarchies of the

the very reason that it violates the national paradigm. They will not see it—it will not be in the worldview they see as the world itself. They do not have eyes to see around the old paradigm. But no reasonable person can doubt, if they spend a minute to think about it, that the global paradigm is consummately practical, the only possible future we may have.

Paradigm shifts occur as anomalies accumulate within the old paradigm. The world will watch as national governments and the clumsy, toothless international organizations they create try to stop nuclear proliferation and climate change. Industrial production will concentrate in the poorest, weakest nations with the worst labor laws and the least environmental enforcement. The effects of irreversible climate chaos will become apparent to everyone. Tropical forests will be "developed" for national economies. Idealistic international protocols will set goals and targets that national governments will be free to violate at will. Nuclear weapons and long-range missiles will be developed by second and third tier nations in the name of national defense. Wars will break out over resources. Populations will soar in countries that can least afford them. Climate refugees will cross borders and overwhelm relief systems. These effects are not predictions. Every one of them—every anomaly that disproves the viability of the national paradigm—has begun already and will continue into the future so long as the old paradigm remains in effect. Nationalism will weaken as the trends continue, until a breaking point is reached.

The breaking point will come when people feel their own lives are at stake. When they see the possibility of collapse before their own eyes, the paradigm will begin to shift.

Global consciousness will be a moving force not when people feel it is reasonable but when they feel it is *essential*, when they know that their survival depends on it absolutely.

They will move when they see the tiger coming.

THE SHIFT FROM THE NATIONAL TO THE GLOBAL PARADIGM will come at the same moment as the shift from the economic to the ecologic. The old paradigms are compatible with one another, but neither is compatible with the future. They will vanish, and a single new paradigm will take their place. Nationalism will live on, as will the economy, but a single global-ecological context will contain them both.

The new paradigms—the ecologic and global—are distinguishable only in terms of the worldviews they replace. Their convergence in the near future is not, therefore, a coincidence; it is an identity revealed in time. The ecologic paradigm depends on a united humanity acting as a whole in relation to nature.

The global paradigm depends on ecological awareness to balance nature with an increasingly powerful humanity. Each depends on the other to the extent of *being* the other. They are a single image of humanity: mature and respectful of itself, in relation to the non-human world. All beings will live under one sky.

The global paradigm and ecologic paradigm are one and the same.

THE QUESTION THAT REMAINS IS: WILL GLOBAL CONSCIOUSNESS arise on its own soon enough to avert the crisis?

The answer, I believe, is no. Modern thinking is too constrained by money and national prejudice for global consciousness to rise quickly enough by itself. It will need help. That is why I believe it is time for civil disobedience on a massive scale.

Civil disobedience, properly exercised, is a tool not of opinion, but of truth—truth that transcends the mind and the intent of the participant. The power of civil disobedience is not to convince, but to show, to point at what is there, whatever it may be. Civil disobedience is consciousness itself, not a configuration within consciousness. It is people looking. By bringing consciousness to an issue, civil disobedience makes people think things they would not normally think. It shakes them up. But it does not ultimately tell them what to think. If thousands of people are arrested in the name of preventing climate change (or some other global cause), people everywhere will think about it. They will have to take another look. If it turns out that climate change is a hoax, an overstatement, or an emotional outlet for the environmentally insane, that will be the truth that is revealed.

Truth, whatever it may be, will not come to the surface quickly enough on its own. It should not. It needs life wanting to live. Life must strive for truth to deserve itself.

The Canadian tar sands and the Keystone XL pipeline are symbols of a larger truth. They are a multibillion dollar investment in the wrong direction. Physically stopping them is doubtful, but physically trying to stop them will help create a new level of human consciousness.

April 18, 2012 – Louisville, KY
High 71°F – Low 46°F – Precipitation 0.0 inches

There is talk of another tar sands pipeline running across Canada and down through Vermont and New Hampshire to a seaport in southern Maine.

NPR's *On Point* devoted its program today to recent weather trends and their connection to global warming. Sixty-nine percent of Americans now believe weather events are made worse by global warming. Tree swallows are nesting nine days earlier, and the annual Christmas bird count shows that bird ranges have shifted four hundred miles north. On the same program, May Boeve, executive director of 350.org, announced the worldwide 350 event "Connect the Dots" for May 5, noting that members in Kentucky would be at the Derby lending creative names to racehorses, names such as Missing Ice Cap, Florida Under, and XL Nightmare. That was one of our ideas ... I guess we will have to do it now.

April 21, 2012 – Louisville, KY
High 52°F – Low 46°F – Precipitation 0.20 inches

Presumptive Republican presidential nominee Mitt Romney said yesterday that he would have no hesitation building the Keystone XL pipeline. "I will build that pipeline if I have to do it myself," Romney vowed before a crowd of state Republican Party leaders in Arizona.

April 23, 2012 – Louisville, KY
High 63°F – Low 39°F – Precipitation 0.0 inches

My local environmental group, 350 Louisville, met this morning to discuss the "Connect the Dots" event on May 5. We are making Derby hats and special T-shirts for the day, and we will be handing out "Derby Dot" stickers with horses' names that relate to climate issues. Some names we came up with are Fossil Fool, Karbonator, Pants on Fire, What the Frack, Too Soon Typhoon, Atmos Fear, Catass Trophe, D. Nyer, Sun Harness, Crispy Crops, and Dirty Tarzans.

April 24, 2012 – Louisville, KY
High 65°F – Low 44°F – Precipitation 0.0 inches

My crew started a new solar installation in Louisville today: Dan and Baron putting panels on the roof, Andy running wire and conduits, and me running back and forth. We should have it fired up tomorrow.

I PARTICIPATED IN A CONFERENCE CALL THIS MORNING (WHILE the crew was overhead banging on the roof) to discuss the possibility of installing solar generating facilities for rural electric cooperatives in Kentucky. This would be a way of graduating from small, residential installations to larger generating facilities, which would feed the grid as a whole. Last fall I sent a letter to the Kentucky Association of Rural Electric Cooperatives offering my services *free* in designing and installing any photovoltaic system up to 12 kilowatts for any rural electric co-op in central Kentucky. The co-op would have to buy the equipment and have their own electrician hook it to their grid, but

I would engineer the system and pay my crew to put it in place. I made this offer because in this area and some other areas of the United States, solar energy remains a dreamy, idealistic, far-off-in-the-future, Al Gore, California hippie proposition that people talk about all day long but that nobody does in the real world when they need to plug in the toaster. It's just something you see in magazines, something a few eccentric characters put on their houses.

Over and over, I hear people say that there just isn't enough sun in Kentucky to make solar "pay off." I remind them that Germany has the highest percentage of photovoltaic production in the world, and that Kentucky has more sunshine *everywhere* than Germany has *anywhere*. But talk is just talk. It doesn't matter what I say or anybody else says, what matters is what solar can do. So let's quit talking and build one, a small one, and I'll help you pay for it. Let's get some hard numbers to look at, and let's make them *your* numbers, not mine. Forget about fossil fuels versus renewables, coal versus solar, good guys versus bad guys. Let's set the ideology aside and look at what is really out there in the world. I'm betting you'll like what you see. The first one's free; you'll get hooked on it, and I'll get my money back when you come begging for more.

I never heard anything back from the Kentucky Association of Rural Electric Cooperatives. No "thank you," no "will take it into consideration," no acknowledgment of receipt. But the letter got around, and I did hear back from a few individuals from two or three co-ops, showing interest.

There were ten or twelve other people on the conference call from environmental groups and electric co-ops. I outlined my offer, disclosed my motives, and answered questions for about an hour. Interest came mainly from the environmental groups. There were some technical questions from the co-ops, but very little warmth toward the idea.

I'm not sure why this is. I could understand reactions such as: "this sounds great, but we don't have funds for this sort of thing," or "the numbers just don't add up from our point of view," or "great idea, let me talk this over with the board." These would be typical "no thanks" reactions to any sort of commercial proposal. Instead I'm getting mostly silence—a "where is this guy coming from?" sort of silence. Maybe it's all in my head. I have had very good discussions with electric co-op representatives and engineers, and genuine interest in renewable energy from some of them, but mostly the things I talk about are not in their world. Co-op people are not opposed to solar energy, they just can't see it this side of the horizon. After an extended discussion about renewable energy one day with a representative from my own electric co-op, he said to me, "Sooner or later we're going to have to do all this ... but I would never ask anyone to change their lifestyle." Material progress comes first.

There are other possible reasons for the cool reception. A great deal of environmental sluggishness is built into the electric generating industry in general and into nonprofit distribution cooperatives in particular. Cost is everything. You can't raise the price of electricity by making it bigger or better or sexier. Electricity is just plain electricity. Even if you make it greener, it still does what it always did. All you can do from a business standpoint is control production costs. Cheaper is better, *period*. This is institutionalized through the Public Service Commission. In Kentucky, as elsewhere, the PSC requires that every utility, profit or co-op, provide power at the least possible cost to consumers. This requirement is classic economic paradigm thinking. If coal is cheaper, you burn coal—end of discussion.

Cost is defined exclusively in dollars. Dead fish, sulfur dioxide in the air, mercury in the water, buried streambeds, eroded mountain topsoil, chronic asthma, and climate change are not money and therefore do not cost anything. If the true expense of coal were internalized in its price—if the coal industry had to pay the social and environmental costs of mining and combustion—solar would be cheaper by far. But *true cost accounting* is classic ecologic paradigm thinking, not the kind of thinking that makes things happen in the "real world."

Distribution co-ops are further hampered by the fact that, as nonprofits, they do not accumulate capital and cannot justify capital expenditures that don't have fairly rapid returns. The cost of solar is all up front. Its operating costs are near zero, and it requires no fuel whatsoever, so you have to pay the entire cost of thirty to fifty years of energy production at the beginning.

So there are reasons why solar is a hard sell for rural Kentucky. But I am reasonably sure that reason has little to do with it. It is a matter of the worldview within which reasoning begins.

April 28, 2012 – Louisville, KY
High 81°F – Low 51°F – Precipitation 0.55 inches

A local news source reported today that a surface mining permit has been issued for tar sands extraction here in central Kentucky. A company will begin mining this summer from a twenty-acre site but plans to expand its operations throughout the region. "We wanted to use this property to show others in the area our operation and how clean it is, so we could begin to look at leasing other land in the area," the company president said. "There is basically no emissions from this process, because we don't use heat." Both the chemicals and the water used in the process will be recycled. "We really think that once this first unit is operational, it will become the technology of the future," he added. The report was upbeat and pro-business with a few dismissive references to environmental impact. The sandstones of south-central Kentucky contain the equivalent of six billion barrels of heavy oil—oil that is now selling at $100 per barrel. Do the math.

April 30, 2012 – Louisville, KY
High 84°F – Low 62°F – Precipitation 0.46 inches

According to the Sierra Club and a CNN report, the United States currently imports more than half a million barrels a day of bitumen from Canadian tar sands. That amount is set to triple to more than 1.5 million barrels by 2020. An additional 3.1 million barrels will be exported every day through a newly planned 10,000-mile pipeline complex, to be built at a cost of almost $40 billion. The Keystone XL is part of the expansion. Other pipelines are proposed to run west across the Rocky Mountains to Canada's west coast, and east through New England to the Atlantic coast. All proposed routes would pass through sensitive ecological areas.

May 10, 2012 – Louisville, KY
High 70°F – Low 50°F – Precipitation 0.0 inches

James Hansen, in an article in the *New York Times*, has some updated figures for the carbon content of the Canadian tar sands. They contain twice the amount of carbon dioxide emitted by global oil use in our entire history. "If we were to fully exploit this new oil source, and continue to burn our conventional oil, gas, and coal supplies, concentrations of carbon dioxide in the atmosphere eventually would reach levels higher than in the Pliocene era, more than 2.5 million years ago, when sea level was at least fifty feet higher than it is now," he said. "That level of heat-trapping gases would assure that the disintegration of the ice sheets would accelerate out of control. Sea levels would rise and destroy coastal cities. Global temperatures would become intolerable. Twenty to fifty percent of the planet's species would be driven to extinction. Civilization would be at risk ...

"The concentration of carbon dioxide in the atmosphere," he continued, "has risen from 280 parts per million to 393 ppm over the last 150 years. The tar sands contain enough carbon—240 gigatons—to add 120 ppm. If we turn to these dirtiest of fuels, instead of finding ways to phase out our addiction to fossil fuels, there is no hope of keeping carbon concentrations below 500 ppm—a level that would, as earth's history shows, leave our children a climate system that is out of their control."

The Practical Realm of Consciousness

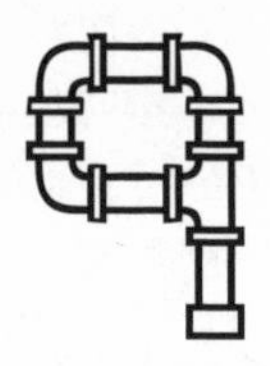

May 14, 2012 – Louisville, KY
High 75°F – Low 61°F – Precipitation Trace

TransCanada has reapplied for the northern leg of the Keystone XL permit. The State Department has agreed to respond to the renewed permit request sometime this year.

I'M A PRACTICAL PERSON. I HAVE TO BE. I RAN MY OWN CONSTRUCTION company for many years, and I still run a small solar installation company. I have to think about what will work and what won't. If I get lost in dreams about the perfect floor plan or the perfect building or the perfect solar array, I'll end up ordering the wrong equipment or paying people to put things together that I will have to pay them to take apart later. To build anything you have to think first; you have to build it in your mind. It begins as imagination, but the real world comes crashing in at some point. To do what you are thinking of doing, you have to mold your thoughts into things that will fit in the real world. You start with anything and everything, then narrow it down to what will work.

There is a particular realm of human awareness where this happens. Raw imagination that has not yet collided with physical reality is molded and shaped to simulate the physical, social, or political world. I call this the *practical realm of consciousness.* It is pure imagination, but practical imagination. It is a special place in the human mind that is both infinitely expansive and severely limited, a dimension larger than the physical world and smaller than the purely imaginary world. Anything can happen in this realm, but very little does. It is this realm of consciousness that makes humanity human, on both the individual and collective level. We draw a blueprint on paper today and live in it tomorrow; we imagine a world today and step into it tomorrow.

Social insects direct their collective behavior through a process known as *stigmergy*. They don't think things through, but they do get things done. One ant with the help of a few friends starts trying to move a piece of food toward the nest. The others climb on top of it, feel it with their antennae, poke it here and there with their legs, then push it along or try to pick it up. Other ants scurry about on all sides, trying to move other things. If the ant and his friends are able to move the object even a little, other ants on all sides drop what they are doing and join in to give a hand. Successful effort attracts attention, and help is automatically recruited. But if the object won't move, other ants ignore it, and eventually those trying to move it give up, too. Beneficial behavior is recognized and rewarded by the group, and the group benefits as a whole. But this is accomplished entirely through trial and error. There is no forethought, no imagining possible outcomes, no practical realm of consciousness. Stigmergy usually works in the long run, but enormous energies are expended in *physically* exploring possibilities before discovering practical possibilities.

Humans engage in stigmergy all the time, particularly in collective behavior. It's how we learn. But once we learn, we save a boatload of time and energy imagining possibilities instead of trying each of them physically. We don't have to try all the wrong ways before we find the right way. More importantly, by extending the practical realm of consciousness beyond established experience, we are able to do things that have never been done before. We think of things with which we have absolutely no experience, and make them work. We do things in our minds before ever doing them in the world. Some things, such as traveling to the moon or managing a planet, can only be done this way. Success has to be known from the start, without practice. There is no stigmergy involved. We can't do everything wrong—or anything wrong—to find what works. Parts of the overall plan can be tested, a rocket engine fired or an experiment run in an enclosure, but the trial ends—and error must end—with liftoff. Thinking has to be right. Very soon we will need this realm of consciousness more than ever before.

So what works? How do we apply the practical realm of consciousness to the environmental crisis in general and to the Keystone XL pipeline in particular?

This is where the paradigm comes in. The paradigm is the worldview: the real world as perceived by the vast majority of human beings at a given time. The practical realm of consciousness is the intersection of imagination with the perceived world—with the paradigm. What is *practical*—what works—will look very different in the prevailing economic and national paradigms than it will look in the global-ecologic paradigm. What does work, what should work, and *for whom* it works depends entirely on who is doing the looking and from where. Practicality is a function of being. Objective facts, such as stars and planets in the night sky, look the same no matter the paradigm, but what pops up in human consciousness depends on the conditioning that is fed into it. If the world is for human consumption, or for Americans, what "we" can do, and should do, will reflect who "we" are, but if the world is for life, what "we" should do will reflect another "we" and will look completely different. The same realm of consciousness will produce a separate picture and a separate set of options. The successful use of the practical realm of consciousness in securing human survival depends less, therefore, on what we decide to do within a particular paradigm than on which paradigm we end up doing it in. We will not know what to do, in other words, until we know who we are.

What I am going to do next is a bit difficult for me: I am going to try to look at the practical realm of consciousness within the economic paradigm. I don't see things this way myself, so I might not get it right, but I will try. I've talked to enough people to have a pretty good idea of what they think, and I'm a businessman. I have a house and a family, and I drive a car, so I'm close enough to the economic paradigm to be able to describe it. I will try to be fair.

We need energy for everything we do. If we do not find a secure and steady supply soon, our way of life as we know it will cease to exist. We need energy to drive cars, heat houses, turn on lights, power computers and telephones, and manufacture all the things we need. If you don't drive a car, the bus or the train or the airplane you ride still needs energy. Even food comes from energy: we need it for tractors on the farm, for trucks to bring produce to cities, and for the fertilizers that make things grow. If we don't do something now about our needs tomorrow, we will freeze in the dark or starve in our own kitchens.

The best way—the only way—to secure a continuous supply of petroleum, coal, and natural gas at a reasonable price is to let the free market do it for us. As the price goes up, the incentive to find more goes up. When free people are allowed to go out and find more and bring it to the market, the price will go down. It's simple supply and demand. Even energy conservation, which we need, is a matter of supply and demand: the more expensive energy becomes, the more money you save using less. You don't have to tell people what to do or not to do. You don't need bureaucracies and agencies and well-meaning people getting in the way of simple market dynamics. Just leave people alone and they will solve problems out of their own self-interest. Call it greed if you like, but when entrepreneurs and consumers do what they do, we are all better off. Business people make profits only if they produce and sell what the public wants to have.

The free market system is the simplest and most efficient way to solve the energy crisis. But it's more than that. It's a matter of freedom, something this country was founded on. As Americans, we want as much freedom as can be provided within the law. We want freedom to use and enjoy property without government pretending to know what is best for us. The beauty of the free market is that it allows us to work for the common good without being told how to do it. Only work that benefits others gets paid. If it does not benefit others, there is no money in it, because nobody will pay you for it. If the company you work for fails to meet market demands, it has the freedom to go out of business, and you have the freedom to go work somewhere else.

The environment is something that has to be taken into consideration. There are some bad apples out there, but most companies are very sensitive about their corporate image and do everything they can to run a clean operation. Nobody wants to be seen as a polluter. The water is a lot cleaner than it used to be, and the air is a lot cleaner, too, mostly because companies are cleaning it up themselves. They will find better, cheaper ways to keep the environment clean if they are not forced to make changes by government agencies that don't understand the problems. An example of good corporate citizenship was the ozone problem we had back in the nineties: chlorofluorocarbons were breaking down ozone in the upper atmosphere and allowing ultraviolet radiation through. It was the companies themselves, the producers of CFCs, that got together in Montreal and agreed to stop production. They found substitute aerosols, kept their factories running and their employees on the job, and saved the general population from skin cancers. Corporations are like people—they want to make a living, but they also want to do the right thing.

As far as the Keystone pipeline is concerned, we just have to have it. Sending crude oil through a pipeline is a hundred times more efficient than sending it by truck or rail, and a pipeline is safer, too. Thousands of people will get jobs building, operating, and maintaining the pipeline, and American consumers will get lower prices at the pump. If we're going to grow, we have to have secure energy at a good price. And we do have to grow. All we have to do is let the free market bring the oil to us. It works by itself, and it's a truly marvelous system.

The environmentalists opposing the pipeline are well meaning, I am sure, but they have no idea what they're talking about. They're way off base on this one. There's the possibility of a spill, true, but nothing is perfectly safe. If you wait until you have everything guaranteed before you start, you never end up doing anything. Maybe the Sand Hills in Nebraska isn't the best route: so change the route! Don't condemn the whole project for one section of the route. And the climate? There's no way to know about that one. Some say it's getting warmer, others say we're due for another ice age. Maybe it is getting warmer, but how do we know we are causing it? And if we don't know if we're causing it, how can we do anything about it? We could bankrupt the economy trying to reverse climate change and then find out we had nothing to do with it in the first place. To base a country's future on the possibility that the climate might be getting warmer and it might be because of fossil fuel, and we might be able to avert it, is awfully fuzzy-headed thinking. All of those people protesting the pipeline will probably be talking about something else a few years from now anyway; we will be glad we didn't listen to them now.

We need the energy, we need the jobs, and we don't need government bureaucracies telling us how to live our lives. This project is privately funded: we can have what we need to live the American lifestyle and have our freedom, too, without any cost to the taxpayer. Is there any real reason to stand in the way?

Whoa! ... I about halfway convinced myself there. But that is the way a paradigm works: it is logical within its own system of logic. It is never right or wrong: right or wrong exist within its limits. A paradigm is a worldview: a picture of how the world looks. What it does not see is not in the world.

Things in the economic world are there for us to use: trees are lumber, soil and mountains are overburden, and people are human resources. Maximum economic growth is the unquestioned assumption of human purpose. Happiness is not money; happiness is *more* money. You're not greedy because you want a better material life for yourself and your family; that's what everyone wants, and has a right to, if he plays by the rules. Continuous and sustained economic growth is not a personal belief, it is an assumption built into the structure of the world. This worldview has been successful for so long because it shows how the marketplace channels individual self-interest into collective interest without effort or any sort of collective intention. It has done so much for so many that it is not surprising to find few people able to see beyond it, or to see the possibility of anything outside it.

But there are two major problems with the free market system. First, while the market mechanism is efficient in turning resources into products and bringing products to consumers, it does not distribute them evenly. Consumption is a factor of wealth and not of need. Rich people get to have more stuff whether they need it or not. People die of starvation and exposure because they are unable to work the system. They don't know how to play the market. But this is a problem of the market, *not* of the economic paradigm as a whole. You can be critical of how the market works in distributing goods and still assume that *having* goods, and having *more* goods, is fundamentally a good thing.

(The word *goods* itself reflects the paradigm.) You can push for more equality than the market offers while assuming that there should be more stuff for everybody. This assumption, as progressive as it may be, remains firmly within the economic paradigm.

The second problem of the free market system is the environment. This is where the economic paradigm runs into a brick wall. The market does not know what life is. When it looks at pigs, it sees pork bellies; when it looks at trees, it sees two-by-fours. There was a time when it even looked at people and saw commodities. That's not because the market is evil, it's because all it can do is follow the money from production to consumption. It doesn't see anything else along the way. We shouldn't ask it to see anything else. We shouldn't ask it to see the difference between a stream and a sewer, or between a forest and a coal pit. The market is good at moving things around from person to person, but it is no good outside of the human community. It can't say what a tropical snail is for, or how much a species of club moss is worth. The market does not value life. The value of life is where the economic paradigm ends and the ecologic begins, and where the practical realm of consciousness shifts from promoting wealth to promoting health.

A SIMILAR SHIFT IN THE PRACTICAL REALM OF CONSCIOUSNESS occurs between the national and global paradigms. What we should do about climate, population, world trade, nuclear weapons, etc., will look very different if "we" are Americans first and people second than it will if we are people first and Americans second. What we do is a function of who we are. From the nationalist perspective, this is what I keep hearing:

Available energy is what keeps this country going and what this country is running out of. We must have a reliable supply for our own needs, or we will end up beholden to some foreign power in the Middle East, South America, or Africa. We have a hundred-year supply of coal, virtually unlimited new reserves of natural gas thanks to new hydraulic fracturing technology, and there's plenty of crude oil offshore and in places such as North Dakota and Alaska. All of this is sitting there waiting for us to use it. It's already ours. Why should we give billions of dollars to unfriendly countries halfway around the world when we have our own resources right here? Why should we spend hundreds of billions of dollars sending the army to secure petroleum supplies in the Middle East when there is a big supply in a friendly country right next door? Sending the army should be a last resort; our own reserves should be a first resort. We have wind, hydro, and solar, too. They do not amount to much, but why not go ahead with them? Every drop of energy we produce here is a drop we will not have to buy overseas. This is money in our pockets, for our people and our companies, and good for our balance of payments. If we want to continue to lead the free world, we cannot depend on anyone else for vital resources. This country was founded on independence and has to remain independent to keep its position in the world.

The American system is the most prosperous on Earth. We have built the best schools and hospitals in the world. We won the Second World War and defeated communism. There has never been a better society than ours and likely never will be. We should take a long, hard look at what we have before we start nitpicking it to death.

There is no way to not build the Keystone XL. There's a huge amount of energy up there, nearly as much as in Saudi Arabia. We're much better off paying Canada than paying people who don't like us and want to destroy us as a society. This is exactly what we need! If we don't buy tar sands oil, the Chinese will, and then where would we be? The Keystone XL has nothing to do with carbon in the atmosphere, it's just a way of getting the oil here instead of somewhere else. The oil will get burned either way. The only thing not to like is that it crosses the whole country from north to south, but when it gets to Houston, the oil is not really ours anymore. It could be sold anywhere. Perhaps Congress should pass an export tariff to keep it here.

Admittedly, there are environmental problems associated with the pipeline and with the tar sands. Canada has its own environmental agencies, so we're not concerned with the tar sands extraction itself. That's within their sovereign territory, and there would be nothing we could do about it in any case. There is no way to eliminate risk; the best we can do is manage it. We have to trust the people building it and trust the inspectors to do their jobs. Properly constructed, this pipeline could be a lifeline to a prosperous future for the people of America.

As far as climate change is concerned, it may be as real as they say it is, but it's as much the fault of countries such as China and India as it is ours. It is ridiculous for them to be exempt from emissions restrictions under the Kyoto Protocol. Why should we pay the price of reducing our emissions while they are free to pollute as much as they want? The Kyoto agreement is not in our national interest, and we should not abide by it.

Clearly, the national paradigm is inappropriate for managing the global environment. Each country is out for itself, and the interests of humanity as a whole are not represented. There are too many parts and not enough whole. National prejudice distorts the perception of global issues and focuses attention instead on injustices among competing countries. Attention flows away from what is actually happening on the global level toward petty bickering between competing nations. This is not where our attention should be. People breathe air without regard to national allegiance, and people interact with the oceans, the forests, and the climate as human beings and not as citizens of any particular country. Yet we have no understanding or control over what we are doing as a whole. We have no means of perceiving ourselves *as humans* in relation to the natural world, and no means to coordinate global action. Within the national paradigm, humanity as a whole has no sense of self, no voice, and no means to do what it needs to do about the problems it faces.

PARADIGMS DO NOT SHIFT EASILY. UNDER ATTACK, THEY struggle to preserve themselves by cramming new information into old categories. New facts that do not fit the prevailing paradigm—what Thomas Kuhn calls *anomalies*—are "shoehorned" in. There is a perceived need to keep the existing worldview alive for as long as possible. To preserve the Earth-centered universe, for instance, explanations had to be found for the *retrograde* motion of the outer planets. The outer planets move westward across the sky most of the time then, strangely, reverse their direction! They go eastward (in relation to the background stars) for a while before switching direction again and heading back westward. This made no sense at all within the Earth-centered worldview. How could God make the outer planets behave in such irregular fashion? There had to be an explanation. The answer given was something called the *epicycle,* a little circular curlicue motion within the larger orbital motion. An epicycle was never observed in planetary motion, of course, but it did wonders to explain retrograde motion within the logic of the geocentric paradigm. It had to be believed in order to keep the earth at the center of the universe.

But a paradigm weakens as anomalies accumulate. As observations improved over the years, astronomers found that one epicycle was not enough to explain what the planets were doing: epicycles within epicycles became necessary. Planetary motion grew complicated and distinctly unheavenly. Explanations became too numerous and forced, and the solar system took on a complicated, untidy look that was unworthy of the hand of God. Ultimately, simplicity could be restored only by the realization that we are *ourselves* on a planet—that is, by shifting the paradigm. Retrograde motion within the new paradigm was how the outer planets *appear* to move as we, riding on the earth, overtake them in our smaller orbit around the sun. Nothing changed in the sky—the planets moved as they always had—what changed was *us*. We no longer understood creation to be radiating out from ourselves. This was the moment of humanity's emergence from childhood to adolescence. We found a larger picture and a place for ourselves within it. Most importantly, with the center of rotation at a distance from the earth, we could picture ourselves from the outside. We were in a space that did not focus on us.

A similar shift is happening now. The economic and national paradigms are entirely logical and consistent within themselves, but anomalies are appearing that reveal their limitations. It is beginning to look as if the world is larger than either of them. Anomalies can still be shoehorned into existing worldviews, but they are accumulating faster than they can be disposed of. It is looking as though the living world does not hold human consumption to be its highest purpose, and as if the universe may not revolve around the nation-state after all.

How, then, would the global-ecologic paradigm take shape within the practical realm of consciousness? Within a rising global consciousness, how would we learn to perceive what is happening, how would we imagine solutions, and how would what we imagine intersect with the real world? How do *we* look from the global, ecologic standpoint? It is not a pretty picture.

We have ravaged the land. We have fouled the air and poisoned the oceans. We are a parasite on the living world. Forests are disappearing, ice caps are melting, and coral reefs are dying. Humans have moved across the land like a plague of locusts. Are we a mold, an infection, a disease on the face of the earth? When I was born, the earth had more than twice as many trees and fewer than half as many people. As I walk through the woods, I often wonder whether the trees, if they had the choice, would risk a nuclear holocaust to be rid of us. It would be hard medicine for them, radiation treatment. Would they risk it?

But this sort of self-loathing goes nowhere. We may have made a mess of things, but we were put on this Earth along with the whales and cockroaches, not by our own volition. We did not choose to exist. We have the right, therefore, to be here. Like a raging teenager, unaware of the world around him as he tries to find out who he is, we have left a wide swath of destruction. But we have the right to be here. The ecologic paradigm by itself does not guarantee our right to exist—we are free to die off, as have most species of earthly life—but it does guarantee our right to *try to exist*. We have the right to envision the world with ourselves in it.

The world is, I believe, the same thing as the worldview. We create in our minds the world we live in. People in the Middle Ages lived in a universe with the earth at its center, and people now live in a universe with human consumption at its center. These are very real places. As the paradigm shifts, we will move into a new world with its own definitions of reality. We should widen our minds to imagine anything and everything, then narrow them to what fits in the physical world. The more we know about the world, the more creatively we will act within it.

The global-ecologic paradigm in the practical realm might look something like this:

The climate is changing. The change is a result of the increase in atmospheric carbon dioxide concentrations since people began burning coal, natural gas, and petroleum on a major scale 150 years ago. If global warming does not progress into the carbon cycle feedback loop and permafrost feedback loop, the climate can be restored to balance over time by reducing carbon emission. There are three measures to be taken immediately: reduction, conversion, and research. Reduction in carbon emissions to near zero by eliminating most carbon-based fossil fuel combustion, conversion to entirely renewable energy sources, and research to enhance understanding of what is happening in the biosphere. If there were time, we would do the research first, determine appropriate atmospheric carbon levels, and only then implement reduction measures. But we do not have that time. We know that the current level is too high, even if we do not know how much too high, therefore we should begin reducing levels immediately. The research will, in time, show us what the appropriate level is.

An equitable way to reduce carbon emissions would be a permitting system based on an annually adjusted per capita "right to pollute." First, determine how much carbon is to be allowed into the atmosphere in a given year and divide that amount by the world's population. The resulting number is each person's carbon allowance for that year. Then, issue carbon emission permits to each national government based on this amount multiplied by its current population. Each country may either (a) emit this much carbon, (b) emit less than this amount and sell excess permits on the open market, or (c) buy permits to emit more. The total global emissions would remain the same whatever individual nations decide. Their decisions will be heavily influenced by the price of carbon permits: the higher the price, the more money they will make by adapting renewable energy sources and selling their permits, or the more money they will spend buying permits from abroad. Permit prices will increase each year as emission levels are reduced and fewer permits issued. This policy will increase incentives for renewable energy installations.

There will be pain and dislocation among older industries as the transition is made, but the system will be equitable and less bureaucratic than mandated levels on an industry-by-industry or nation-by-nation basis. Everyone will have the same right and the same responsibility. Each nation will have the right to distribute its permits internally as it sees fit. Industries will be responding to market forces rather than to arbitrary government regulation. In time, very few permits will be issued, and the price will be too high for routine combustion of any fossil fuels.

There will be at least two major side effects of such a system: one positive and the other negative. The positive effect, from the global perspective, will be that developing countries with low emissions will benefit financially. There will be a major transfer of wealth from the industrialized countries that caused the onset of climate change to the countries that suffer from it most. From the nationalist perspective, this will be difficult to accept. Why should industrialized countries have to buy billions of dollars' worth of permits from people who have no intention of using them? This proposed carbon permitting system will become acceptable only when the atmosphere is understood to be a limited resource and a global commons. The transfer of wealth will be a welcome side effect of the system when large amounts of new money become available to developing countries for investment in long-term economic growth based on renewable energy.

The negative side effect of a system based on population is that it creates an incentive for population growth. Nations will want to keep populations high to claim more annual permits. This effect will call for another globally based system to reduce population growth throughout the world.

Systems similar to the carbon permitting system will be used to control and allocate other global resources, including minerals, timber, and water. Market forces will thereby allocate resources based on equity and environmental stability rather than only on price. Worldwide environmental protection laws will create an even playing field for industrial production, keeping it from concentrating in areas without protection. Global legislation will shield indigenous peoples from economic encroachment and create biodiversity preserves throughout the earth's major bioregions. Uniform labor laws will protect jobs from exportation to countries without minimum wages, worker's compensation, and collective bargaining rights. Uniform trade laws will protect business from unfair tariffs and national subsidies. Humanity will act as a united world to solve world problems that can be solved in no other way.

The global-ecologic paradigm has its practical logic. From the global perspective, it is much easier than from the economic or national perspectives to understand environmental, economic, and security problems, and much easier to imagine solutions that would converge with real-world conditions. There can be no doubt that the global-economic paradigm is far better suited for the future than either of the older paradigms. But the global-economic paradigm has a drawback that the other two do not have: it does not exist. It is rational, simple, logically consistent, and consistent with the practical realm of consciousness, but ultimately impractical, because the world it inhabits has not yet come into being. It is a worldview without a world.

Bringing the new paradigm to life is more important now than advocating specific practical measures. Confronting climate change within the old paradigms, for instance, would be as likely of success as exploring outer space in the medieval universe.

May 18, 2012 – Louisville, KY
High 84°F – Low 54°F – Precipitation 0.0 inches

The U.S. House of Representatives passed a resolution to approve the Keystone XL pipeline as part of an unrelated transportation bill. The Senate and the president are unlikely to approve this version of the transportation bill with the amendment attached, but there is so much money behind the pipeline that it will eventually find its way through the approval process.

May 26, 2012 – Louisville, KY
High 91°F – Low 70°F – Precipitation 0.0 inches

The International Energy Agency reported that CO_2 emissions reached an all-time high in 2011 and that chances were diminishing for keeping the world temperature rise below 2 degrees Celsius. NASA climatologist James Hansen said that even a two degree rise would be "a prescription for disaster. [But] when I look at this data, the trend is perfectly in line with a temperature increase of six degrees [by 2050], which would have devastating consequences for the planet."

May 31, 2012 – Louisville, KY
High 84°F – Low 58°F – Precipitation 1.02 inches

The level of carbon dioxide in the atmosphere has passed 400 parts per million in some places, according to a report by Seth Borenstein of the Associated Press. The International Energy Agency also announced that global carbon dioxide emissions from fossil fuels hit a record high of 34.8 billion tons in 2011, up 3.2 percent. Environment 360 at Yale University says CO_2 levels are the highest in eight hundred thousand years.

Fracking

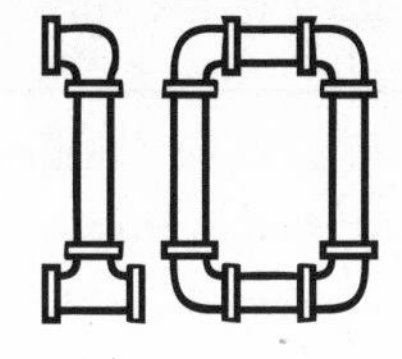

June 8, 2012 – Louisville, KY
High 86°F – Low 59°F – Precipitation 0.0 inches

Common Dreams reports a crude oil pipeline spill in Alberta: Plains Midstream Canada has announced that a large oil spill has erupted from its Plains' Rangeland Pipeline operations in West-Central Alberta. The company estimates up to three thousand barrels of "sour crude" oil has leaked into a large river system surrounded by "pristine wilderness." The spill is the second in thirteen months from Plains Midstream, which has now halted operations as emergency crews attempt to clean up the spill. Plains Midstream is still in the process of cleaning up last year's spill of twenty-eight thousand barrels in northern Alberta. The crude spilled near a native community in late April last year.

FRACKING IS A NEW GAS-DRILLING TECHNIQUE THAT HAS revolutionized the fossil fuel industry. It has increased productivity of natural gas wells and brought prices down in recent years to the point where gas is beginning to replace coal as the major fuel of American electricity generation. Gas burns cleaner than coal. It produces less local air pollution in the form of particulates, mercury, selenium, and sulfur dioxide; leaves virtually no ash; and emits less climate-busting carbon dioxide. Where coal is virtually all carbon, the main ingredient of natural gas, methane, is carbon plus hydrogen. Each methane molecule is a carbon atom surrounded by four hydrogen atoms. Part of the energy released in combustion still comes from the carbon, but the rest comes from oxidizing the hydrogen into water (H_2O), an environmentally harmless byproduct. Natural gas as a source of electricity generation means cleaner air near power plants and less overall carbon in the atmosphere. Gas is still a fossil fuel, however, and therefore not a long-term source of sustainable energy.

Substituting natural gas for coal is an environmental winner all around, at least in the short term. It takes the pressure off the mountains of Appalachia, allows trees and wildflowers to grow on the hillsides, gives streams the chance to run freely, and allows birds to sing in the forest. But substituting gas for coal is not so simple. The fracking process of natural gas *extraction*—the reason gas is now so cheap—is an environmental nightmare all on its own.

Natural gas deposits usually occur within the pores of sedimentary rock far below the surface, with layers of impervious rock above and below. The impervious layers keep the gas contained in a watertight chamber. Fracking (formally known as *hydraulic fracturing*) pumps millions of gallons of high-pressure, chemically laden water down a vertical well casing and then (usually) horizontally into the chamber. Water pressure from pump trucks above ground cracks the rock and releases the gas. Sand particles or other "proppants" mixed in with the water and chemicals move into the cracks between rock fragments and keep them open long enough to extract the gas. Chemicals are added to the water to maximize proper placement of proppants between rock fragments and keep them from retracting as the water retracts. The average fracking well uses up to a hundred thousand gallons of chemicals over its lifetime. Nobody knows for sure what these chemicals do or where they go once they are squeezed into the ground. Nobody knows how long they will remain there. *We don't even know what they are.*

In 2011, the U.S. House of Representatives investigated the chemicals used in fracking and found that, of 2,500 products used, 650 were known carcinogens or listed as hazardous air pollutants. Of these, 279 had at least one component that was listed as "proprietary" or a "trade secret," meaning that fracking companies had no legal obligation to tell federal, state, or local regulating authorities what they were pumping into the ground and, potentially, into the drinking water of nearby wells. The House report concluded that "companies are injecting fluids containing unknown chemicals about which they may have limited understanding of the potential risks posed to human health and the environment."[19] People living near fracking wells do not know what they may be drinking. To provide some legal protection, residents may, at their own expense, hire a laboratory to test their drinking water, but they must do so *before* fracking begins in order to establish a baseline comparison with samples tested after fracking. If extracting companies don't tell them what chemicals will be used, residents have no way to know which ones to test for, and thus no way to prove environmental contamination.

Another study in 2011 identified 632 chemicals used in natural gas operations, many of which could affect skin, eyes, respiratory and gastrointestinal systems, the brain and nervous system, immune and cardiovascular systems, the kidneys, and the endocrine system. Of these, 160 were carcinogens and mutagens (agents that cause genetic mutation). The study recommended that fracking's exemption from regulation under the U.S. Safe Drinking Water Act be rescinded.[20]

Gas drilling companies are way out ahead of the public in the environmental battle over fracking. They knew it was coming and prepared accordingly. Anticipating a public relations disaster, they formed associations such as America's Natural Gas Alliance "to promote the economic, environmental, and national security benefits of greater use of clean, abundant, domestic natural gas." Print, radio, and television advertisements speak in phrases such as "the clean energy landscape," "this clean energy resource," "improving air quality," "adding jobs," "enhancing our energy security," and a "new age of natural gas abundance." But the gas industry is most politically active on the state and local levels, where laws that affect them are made. For the past few years, they have pumped high-pressure toxic dollars into state houses around the United States and extracted multi-megaton concessions before the general public knew what was happening.

In Pennsylvania, for example, they pushed a bill through the legislature that *requires* all municipalities to allow gas well drilling and wastewater pits in all zoning districts, *including residential.* Local governments are not allowed to limit the hours of operation of gas drilling activities, and pipelines cannot be disallowed anywhere.[21] Doctors in Pennsylvania can access the list of chemicals in hydraulic fracturing fluid in emergency situations only, and they are *forbidden by law* from discussing what they have learned about these chemicals with their patients.[22]

Another big problem with fracking is that a lot of methane escapes into the atmosphere during the drilling process, somewhere between 3 percent and 7 percent of the total. That's a lot of gas, and methane is a much more potent greenhouse gas than carbon dioxide—about twenty times worse. *Burning* natural gas may be better environmentally than burning coal or petroleum, but *extracting* natural gas through fracking is worse than coal mining or oil drilling. Between 30 percent and 100 percent more methane escapes from fracking wells than from traditional gas wells.

But what does fracking have to do with the Keystone XL pipeline? Fracking is distinct in technology, geography, fuel type, and end use. The fight against it involves a different level of government, different kinds of people and places, and a different set of energy companies than the fight against the pipeline. Fracking is thousands of separate operations at scattered localities rather than a single, linear, transcontinental project. The struggle against the pipeline, were it to take on fracking as well, might be dissipated by so many separate battles. But they are all part of the same war. A carbon atom from a gas well in rural Pennsylvania is no different from a carbon atom from a strip mine in Alberta or eastern Kentucky. The economic force pumping millions of gallons of toxic chemicals into Pennsylvania and Ohio is the same force digging billions of tons of tar sand from the boreal forests of Canada and blowing the tops off mountains in Kentucky and West Virginia. We must choose our battles carefully, but we must also coordinate them to win the overall war. The fossil fuel industries would like nothing better than to deal with scattered, isolated, local movements throughout the United States, and throughout the world, rather than to deal with a unified, worldwide vision of a renewable energy future. They need not divide to conquer if we are already divided.

June 17, 2012 – Columbus, OH
High 84°F – Low 70°F – Precipitation 0.32 inches

I had the chance to talk today with Bill McKibben at an anti-fracking action in Columbus, Ohio. He is cautious about getting lost in single issues, whether pipelines, coal mines, or fracking wells. "We can't fight this one pipeline at a time or one gas well at a time," he said. "There are too many of them." His current emphasis is on repealing the billions of dollars of tax subsidies currently enjoyed by the fossil fuel industry: "That way we hit them all at the same time." I argued for the pipeline as a symbol for the fossil fuel industry as a whole and for coordinated actions along the route during construction. "This is the environmental issue of our time," I assured him. But I will know more about how seriously people are taking the issue when I travel the route of the pipeline next month. I will stay in touch with him. His organization, 350.org, supports local groups and local issues such as the one here in Columbus, and it will no doubt support local actions against the pipeline. It's a question of where to commit forces, and whether we should be asking hundreds of young people to begin their careers with an arrest record. "I figure, at my age, what do I have to lose," Bill quipped. "But we should think twice about making cannon fodder of our young people." I agree. Civil disobedience should be used carefully and sparingly. Perhaps we should fill in the front lines with specially formed "geezer brigades."

Bill is off to Istanbul tomorrow, and then on to the Climate Summit in Rio.

June 29, 2012 – Louisville, KY
High 105°F – Low 78°F – Precipitation 0.0 inches

The temperature hit 105 degrees today in Louisville: a new record for the date. The heat is expected to continue through large sections of the United States for another week or so. Is this the new climate, here to stay? Or is it just another routine heat wave? If a really scary new climate pattern were to begin right now, this is how the first day would look. But things will no doubt cool down again and seem normal, for a while.

I have to admit, however, in a twist of intellectual honesty, that I am glad to see all the new temperature records this year. I am not glad to see the beginning of the suffering climate change will cause, but I am glad that it is becoming visible and glad it is happening now. It is much easier to point to a visible tiger. You don't have to convince anyone it is there if they can see it for themselves. Creating pictures with numbers, chemical equations, and inflated alarmist projections is a lot more work than stepping outside and feeling the heat. It will be easier to point to the pipeline and say, "This will make you even hotter than you already are."

Congress passed the transportation bill today without the mandate that would have required approval of the Keystone XL pipeline within thirty days. Republicans in a House-Senate conference committee gave up on their attempt to circumvent the president's decision against the pipeline, but TransCanada has already begun reapplication procedures.

I HAVE ALWAYS BEEN IMPRESSED BY THE COINCIDENCE BETWEEN the ill effects of petroleum and its limited global supply. I have wondered, over the years, why it is that we seem to be running out of oil at about the same pace that it is altering the climate. This seems a fortunate coincidence. With growing scarcity, the price of petroleum will rise naturally, favoring renewable energy sources in the marketplace just as they become essential for environmental survival. A sort of economic gravity will pull reluctant consumers away from oil and toward a sustainable future, whether or not they understand what is happening to the environment. The coincidence suggests a divine plan. The earth is a cosmic seed; we are the growing plant embryo, provided with just enough sustenance within the skin of the seed to power civilization from industrial infancy to a mature, sustainable self-sufficiency. We will spread our foliage into the sunlight and make our own way in the world just as the energy stored within the earth runs out.

I still like the way this sounds, but I'm no longer sure it's true. According to a book I finished reading today, there may be a lot more oil in the ground than we will need to ruin the climate. According to Leonardo Maugeri's *The Age of Oil: The Mythology, History, and Future of the World's Most Controversial Resource*, proven oil reserves are now greater than they have ever been. There is, of course, no more oil in the ground than there was before, but we keep finding new oil fields and, more importantly, keep finding new ways to extract from old oil fields. Proven reserves are known quantities of crude oil that can be extracted through current technological means, so if the rate of improvement in extraction exceeds the rate of oil depletion, proven reserves *increase* over the years. That seems to be what is happening now. The *life index* of world reserves is the ratio between proven reserves and current consumption, which estimates how many more years of oil we can expect. In 1948, the life index was 20.5 years; in 1973, it grew to 32.7 years; and in 2005, to 38 years (which will get us to about the year 2050). But this is only for proven reserves. There is a lot more oil that cannot be extracted easily with existing technology, and there are new oil fields yet to be discovered. The world's *proven* oil reserves are now between 1.1 and 1.2 trillion barrels (65 percent of it in Persian Gulf states), but the U.S. Geological Survey in 1996 estimated total planetary "original oil in place" to be up to seven trillion barrels, of which we have consumed less than one trillion to date. That leaves six trillion barrels to go. If these figures are right, and we keep getting better at extracting oil, we will have another two hundred years of petroleum at current rates of consumption. This does not include another estimated *four trillion barrels* of "unconventional oils," such as the Canadian tar sands.[23]

I don't know if those numbers are right, but I'm beginning to think that large petroleum reserves, vast stocks of unmined coal, and the fracking boom in natural gas extraction mean that fossil fuel scarcity will not be what saves us from climate disaster. Economic gravity will not pull us where we need to go, and the amount of fossil fuel stored in the earth does not appear to be calculated to keep us from destroying the climate. There may be no divine plan after all; we may have to think up something that is not thought up for us. If the cheapest option is to destroy ourselves, we can save ourselves only through deliberate actions inspired by a new dynamic of collective awareness.

Then again, this may be another sort of divine plan, one in which we are not led by the nose. The plan may not be to shove us in the right direction with economic incentives, but to allow us to inspire ourselves with an expanded vision that transcends economic incentives. Perhaps the whole idea is to let us choose how to be. If we cannot escape the economic paradigm—if we choose a vision forever limited by money—that is what humanity is, and that is where it will end. If we choose life over money, we will evolve into an entirely new form of civilization that we can only now begin to imagine. If God leaves the planning to us, we are less likely to create Him in our own image.

July 8, 2012 – Louisville, KY
High 97°F – Low 73°F – Precipitation Trace

Tomorrow I am off to the Great Plains, driving the route of the Keystone XL from the Dakotas to Texas.

PART

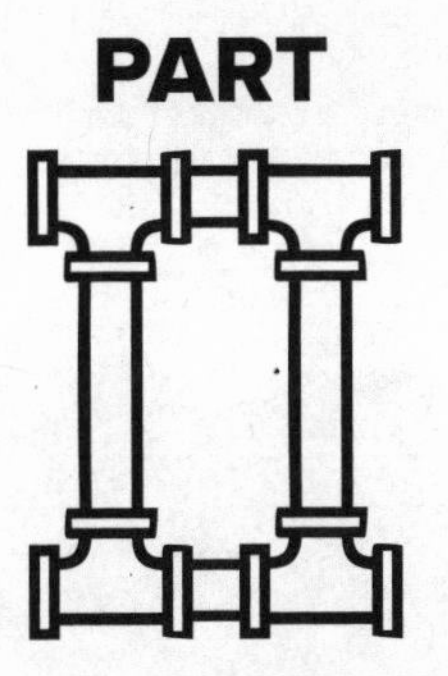

The Pipeline

A sign stands along the path of the Keystone 1 pipeline.

The Dakotas

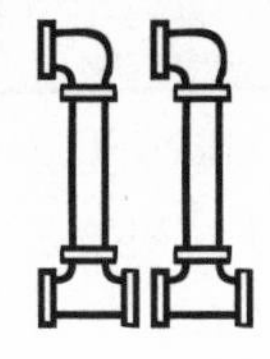

July 12, 2012 – Watertown, SD
High 88°F – Low 67°F – Precipitation 0.03 inches

Last night as I camped on the shore of Pelican Lake, a bit of rain fell on this thirsty ground—hardly enough to settle the dust. I'm in South Dakota, and it's not as hot or dry here as back home in Kentucky, or in the states I have driven through the past few days. The corn is shriveled in Indiana and worse in Illinois, with brown spots in some fields. Market reports tell of soybeans at record prices and corn selling for more than seven dollars a bushel. Crop reports list high percentages of "poor" and "very poor" in several Midwest states. But things looked better in Iowa and Minnesota. Iowa is its usual summer ocean of corn, just beginning to tassel, and it looks fairly good for a bad year. In eastern North Dakota, where the drought has not reached, the cornfields are a dark healthy green. A man with a quarter section of healthy corn can make some real money in a year like this.

I SPENT THE DAY YESTERDAY WITH A FARMING FAMILY IN Cogswell, North Dakota: Paul and Tammy Mathews and their teenage son, Elijah. They have a beautiful new home on a low hill overlooking two thousand acres of corn and soybeans—and, to their chagrin, overlooking the TransCanada Keystone 1 pipeline. The pipe itself is four feet below ground, but the scars it has left on their lives remain on the surface. This is a thirty-inch tar sands pipeline built in 2008. Its diameter is six inches smaller than the proposed Keystone XL, but the patterns of land acquisition, construction, and operation are the same. It does not reach a deepwater port, but it is already pumping Canadian tar sands oil into the American economy. It showed up one day at the Mathews' home like a bolt of lightning. An agent for TransCanada came to their front door in early 2007, without notice, and spread out a map on their table showing the Keystone 1 pipeline going right through their living room!

"We didn't know anything about pipelines, crude oil, or tar sands." Paul said. "We were shaken. They took us totally by surprise. They wanted us to sign right away, right then. People who didn't sign were threatened with eminent domain. They acted like there was no choice—it was going to happen anyway, so why go to all the trouble of resisting?" At first, they agreed to move the right-of-way from through the Mathews' house to 150 feet away. North Dakota law requires a pipeline right-of-way to be at least 500 feet from a house, so the company offered Paul and Tammy money to sign a waiver. They didn't sign. "When I think of pipeline rupture, I think of a flame shooting fifty feet into the air," Paul said. "The thought of that next to my house kept me up at night." Sitting there listening to Paul, I wondered why the state would allow a waiver if their own law required 500 feet. The company threatened condemnation, but Paul and Tammy held out, and the pipeline ended up a full 1,500 feet away from their house.

But the issue is not over for them. The water table is high in this part of the state, and chemicals leaking from a pipeline can travel long distances before seeping into well water. "Every time we draw water out of the faucet we wonder, 'Is there benzene in it or isn't there?'" Paul said. "That's a fear they have thrust on us. I belong to a communal society in North Dakota that respects other people as equals: you're a human, just like me, we can handshake, and you wouldn't put me in such a position just because you have the power to do so. Corporate greed, that's a whole new realm for me to deal with. But let me tell you, there *is* such a thing as corporate greed. I didn't know what that meant before. I'm God-fearing, and when I go to heaven, I don't expect to see any of these major corporations there."

When I asked Paul what it was like to suddenly become a critic of corporate America, he replied that he considered himself something of an "accidental activist." "But who was more of an activist than Christ?" Paul asked, as if surprising himself with the question. When all the dust settled, he and Tammy put a voice to what had happened. They testified at a Nebraska legislative session, and later on in Montana.

But Paul is lukewarm on environmentalism in general. "Unfortunately, I don't think environmentalism is going to work in American society where we judge everything in a financial decision mode. There's not enough money behind it," he said. The other problem he sees with environmentalism is the environmentalists themselves: they get too wrapped up in extreme issues, such as saving snails and lizards. "This is where environmentalism can get injured. If you're trying to range out too far you can get attacked. When you start talking about big things like the climate, that's too far from my reality. I don't want to think about it."

Tammy walked in at that point, and I asked her about the pipeline.

"We just felt we needed to speak out," she said. "We couldn't just lie down and let it happen. That's what TransCanada wanted us to do: just sign the papers. They wouldn't listen to us. How can they do this to us? They were telling us how everything was going to happen. We had no voice; they didn't listen to anything we said, and we were supposed to take whatever they gave us. We decided to stand up. It was all planned out: they picked up the senior citizens and absentee landowners first. Then they could say 'All your neighbors have already signed. You might as well, too.' "

I sensed Tammy had a slightly different take than Paul on what had happened. "Has this all changed how you understand things like climate change and environmental protection?" I asked.

"When we first got into this," she answered, "we were thinking about our rights as property owners. Then, after we got into it for a while, we started thinking, 'What is flowing through this pipeline?' Then I heard about what is happening to the land in Alberta. I've become a lot more aware now. There really is a *huge* planet issue and a fossil fuel consumption issue that we have to face in this country. We would have paid no attention to any of this without the pipeline."

"Tammy took us all to Washington," Paul added. "We went to a protest last November and stood in front of the White House."

"Bill McKibben's organization, 350.org, was doing this civil disobedience thing last August," Tammy said. "I had a lot of 'get-go' in me then. I thought that would be great to go to, but the timing was all wrong for us in August. So we went in November, with our son, Elijah, to surround the White House, reminding Obama what his campaign promises were on the environment. That was a big event for us: thousands of people were there. But I felt like I missed an opportunity not going in August. I've never been arrested for anything. I'm kind of a quiet, shy person, but I really wanted to do it. I wanted our son to experience the feeling of putting yourself out there for something you believe in."

I complimented her for going that far out of her way to show the world what she was beginning to see. But there was a dark side to her growing awareness. "The oil and gas companies are so powerful and influential: I kind of think they're running the world," she said. "I have read books that predict an apocalypse. It's sort of prophetic. It's a very dark future. I think there will be an apocalypse." I told her that it looks that way to me sometimes, but that I try to see the future as a challenge and an opportunity for creativity.

AFTER LUNCH, PAUL AND I DROVE A FEW MILES DOWN THE road to visit a neighbor, Bob Banderet. One morning last spring, Bob saw for himself what Paul and Tammy most feared. "I saw it first," Bob's daughter Meagan piped in. She had been out early in the morning, bleary eyed, feeding her calf. "I saw this thing spraying up over the trees. I knew what it was," she told me.

"There was no mistaking what it was," Bob added. "Those cottonwood trees are a mile and a half away, just behind the pumping station. It was way up above them. I would have guessed it was a stream about the size of a fence post, but it was a three-quarter-inch fitting that had ruptured. They later said the pipeline was pumping 1,100 pounds of pressure. I called the emergency number and told them what I saw, and they put me on hold for about four minutes. That was the longest four minutes in my life."

It's a good thing Meagan and Bob caught it as soon as they did. The line had been leaking for about half an hour, and it took ten minutes to shut the pumps down. "We heard the pumps stop. We can hear them from here. But the plume of oil kept spurting for ten minutes after that."

About twenty thousand gallons leaked into the containment area at the pumping station and onto an adjoining field. A small amount flowed into a pond nearby. The first cleanup crew arrived five hours later.

"There was not enough of a pressure drop for them to detect electronically," Bob said. "Somebody had to see it, and there's nobody there most of the time. If we hadn't seen it when we did, it would have leaked for hours, maybe days or weeks. This could easily have been a major disaster. People don't realize how fragile this pipeline is."

The spillage added fuel to Bob and Paul's concern about what the pipeline could do to their property and their homes. "I am changing my mind," Bob said, after a pause. "I probably wasn't a big global warming person before. Now I'm starting to consider the possibility a lot more. It's changed me. Even in the local community, some of our neighbors look at Paul and me and start saying, 'Oh, they're talking pipeline again,' but this whole thing has changed how I understand what is going on."

Bob added, "Our story is going to mirror what happens on the Keystone XL."

As we finished up, I asked both Bob and Paul how they would react to the possibility of civil disobedience actions along the pipeline. "It won't happen in North Dakota," I assured them, "but it may happen in Texas and Nebraska." There was a long North Dakota pause.

"It wouldn't be me," Bob said. "I would admire their tenacity, and it would depend on what they were doing, but I can't see them doing it for air quality or the climate. I could see it if they were defending their homes."

"What if they were defending their home in the larger sense?" I asked.

Paul thought for a minute. "For the general population, getting arrested because it's getting hot out might not go over well. Bob and I might appreciate that kind of effort, but our neighbors likely would not."

I was not a bit surprised by this. When you're talking climate, you're talking worldview. You're talking big picture. The big picture comes from church, from community and upbringing. There are always politicians, preachers, tree huggers, and talk show hosts trying to get in on your worldview, for their own reasons—some for votes and profit, some for do-gooder causes. But you don't want to get pulled back and forth by the latest fashions in religion, science, and political ideology. You don't change your worldview if the world you see around you does not change; you don't take anybody else's word for what is real. You have to see it for yourself. Without a full vision of the purpose of human life on Earth, changing a worldview can be a very scary, dark, and threatening proposition. As a society, we do not now have a full vision of the purpose of human life on Earth.

July 13, 2012 – Watertown, SD
High 98°F – Low 63°F – Precipitation 0.0 inches

Driving through the wheat fields of northern South Dakota today, I was surfing through the radio band trying to find something worth listening to. I've been taking the back roads, avoiding interstates, and avoiding my usual listening habit (NPR). A preacher on a Christian station was describing a gathering of "over a thousand" young Muslim-Americans somewhere in Wisconsin. "This is a religion that teaches that Allah is the only God and that God has no son. It demands so much of one's individual life and commitment that it persuades young people to strap explosives to their body and detonate them in a crowd."

I DROVE SOUTHWEST ACROSS THE MISSOURI RIVER INTO THE prairie lands of South Dakota. It's much drier here than in the eastern Dakotas. There's more wheat, hay, sorghum, and pasture here, and less corn. John Harter, a landowner near Colome, has several hundred acres of each. The Keystone XL will cross through the middle of his homestead acreage along the northern edge of the Ogallala aquifer.

"This place has a lot of meaning for me," John said. "See that old homestead there?" A ruined stone building lay open to the weather. "My brother and I

John Harter stands by the Keystone XL's right of way.

used to fight over who would sleep next to the wall. It was so cool in the summertime next to that stone.

"My father had this land, and my grandfather before him," John continued. "We've had it since the thirties. If that pipeline leaks on my property, it's done. It's junk. Once that oil gets into this sandy soil and starts moving, I've got nothing left. The land is worth around $280,000, and they wonder why I won't sign off on it for $13,000! I've asked bankers, 'Would *you* do that?' They said, 'No, it doesn't sound like a good deal to me.' But none of them will stand up for me, either.

"If the pipeline goes through, the outcome of this is going to get scary. When it starts spilling and leaking and people start getting sick … It seems like a dark hole I'm looking from, but we know these pipelines leak, and they're gonna leak, and they know they're going to leak. We've put everything we've got into this place, and now somebody is coming to take it away from us. In my mind they might as well be taking it away, because to me, it's going to be worthless," he said.

John has a strong desire to preserve the land and water, and to pass down to the next generation what has been given him. He wants to keep it in the family, and keep it healthy and beautiful. He also has a strong sense of living in the heartland of America. "A big share of our community is built on rural people," he said. "The people in rural America have been a pawn of the government for a long time. The Keystone people are trying to make this an energy security issue. But I think it's a security risk. They're running this pipeline down the center of the U.S. This is our stronghold area; this is our safety zone. We're fifteen hundred miles from every coast. We've got water, we can raise food, and they want to put what I consider a terrorist target out here. Water and food—those are the two things we need to live, so I think this is a matter of national security."

John knows the climate has changed, but he wasn't ready to blame it on fossil fuel consumption. "I know how the climate has been just since I was a kid. We used to have five-, six-, ten-foot snowbanks every winter. This year we didn't have any. I think we are in climate change. I'm not a hundred percent sure what's causing it. I'm kind of skeptical about whether these scientists are honest. I don't know if anybody is honest anymore, after what I've been through with these TransCanada people. It takes your faith out of having a decent system," he said.

I asked him if he gets any support from the community. "Some people agree with me, but they won't stand with me. I kind of understand—if this wasn't going through my land I wouldn't be so worked up about it." When I asked him what he's going to do about it, he answered, "They're taking me through eminent domain court in November. You can't go in thinking you're going to win, but you've got to give it your best shot. What really gets me is, I own this land, I pay taxes on it, and I have to prove that they don't have the right to it, instead of *them* having to prove that they *do* have the right to it. The burden of proof is on me!" Then I asked him about civil disobedience. "When you have all your rights taken away from you, you don't have a lot of power. They're going to have to put me in jail. Civil disobedience is basically what I'm doing now. I got people that will go out there with me, mostly native Americans," he said.

John was deeply wounded by this sudden intrusion into his world. "The worst thing I've ever had happen to me is watching my mother die from cancer, but this is worse in a way, because there was an ending to that," he said. John's eyes moisened as he spoke of his mother. He cares, and he does not want to be left alone in defense of what has always been his.

"They will be judged for this sin when their judgment day comes," John said. "I truly believe that, because it would drive you nuts if you didn't have something bigger than yourself to get you through this. Who knows what God's plan is ... This paints a blacker picture for me."

"I think this is a time in human history when we are being tested," I said. "I don't think there's a plan to the test. We're not going to fit into something preconceived. We have minds; we have spirit; we have eyes to see."

"That's why God gave us free will."

"Exactly. The plan is for us to make the plan."

Nebraska

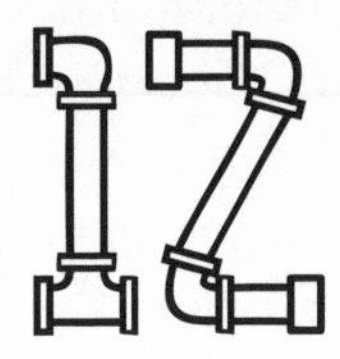

July 16, 2012 – Lincoln, NE
High 95°F – Low 69°F – Precipitation 0.0 inches

One thousand and sixteen counties in twenty-six states were declared a natural disaster last Thursday by the U.S. Department of Agriculture. Parts of the Midwest have seen the driest conditions since 1988, but what distinguishes this drought is its geographic expanse: more than half the entire country, mostly in the southern states. Of the areas I have driven through on this trip, only Indiana and Illinois are in the declared area. Nebraska is not. The wheat crop here was in before the drought, and the corn looks pretty good where irrigated, but the hay looks like amber waves of grain. (It's not grain, it's grass, and it's supposed to be green.) Crossing the Platte River bed two days ago, I noticed it was completely dry. The state has banned any further irrigation from surface waters, and there is no significant rainfall in the forecast.

A recent report from the National Oceanic and Atmospheric Administration ties extreme weather events such as prolonged heat waves and drought to the broader implications of climate change. The past twelve months have been the warmest on record in the United States since the National Climatic Data Center began recording temperatures in 1895.[24] I will be driving into some of the states most affected by the drought later this week and next week.

I MENTIONED PREVIOUSLY THAT I BELONG TO A SMALL, informal group in Louisville that participates in, and often initiates, climate-related actions on the local, national, and global levels. We have no dues, no membership list, no staff, no budget, and no regular meetings or programs. We just do things as they come up. We called ourselves the Louisville Earth Affinity Group (LEAG) for a while, but ended up with 350 Louisville, as many of our actions coordinate closely with the global organization 350.org. In 2009, a number of us went to Washington, D.C., to shut down the coal-fired Capitol Power Plant, and later that year, as part of a global event, organized a large human formation on the Great Lawn in downtown Louisville spelling out the numerals "350." On October 10, 2010, we put together a renewable energy workday that included a solar installation that I directed, and in 2011 we organized *Moving Planet*, another think global/act local 350.org event. We also attend hearings on mountaintop removal and coal ash pollution.

Our group's informal approach avoids the burden of organizational bureaucracy, and it allows us to enjoy each other while we are making a difference in the world. We like being together. Sometimes we just party. Being together and knowing each other well serves an important purpose in building solidarity when we take action. Last August, when we went to Washington for the White House sit-in, we gave each other emotional support as we stepped into the great unknown of bodily arrest. We are teachers, musicians, dentists, housewives, carpenters, nutritionists, solar installers, and nurses. Some of us are old enough to be retired. We are not hardcore politicos. Individually, we belong to the Sierra Club, Kentuckians for the Commonwealth, the National Audubon Society, MoveOn, and the Kentucky Solar Energy Society. We are ordinary people who sense the earth slipping away beneath our feet and who feel compelled to take action. We have hope.

There are a lot of ways to organize, but I think ours is an especially good model for environmental action in the coming years. It is nearly identical to what a group in Lincoln, Nebraska, is doing. They held a meeting while I was in town. They have no official name and usually refer to themselves as "the coalition," "the group," or "350." A dozen or so people showed up at Mary Pipher's house with snacks, wine bottles, and covered dishes. After half an hour of small talk, Adam Hintz reported on his outreach film presentations in small towns around Lincoln; Aubrey Streit-Krug told of her work helping with Mary's writing; Ken Winston spoke of the latest on the legislative front; I reported on my experiences in the Dakotas; and Mary read the introduction from her new book, *The Green Boat: Reviving Ourselves in Our Capsized Culture*, due to come out just after this one. (After you read this book, go and buy hers, too.) Her book is about the dynamics of this small group and the larger coalition that includes Bold Nebraska, the Sierra Club, the Nebraska Wildlife Federation, Audubon Nebraska, Nebraska Farmers Union, Nebraskans for Peace, and university environmental groups. It chronicles their struggles with the Keystone XL pipeline. Mary has written several other important books, including the bestseller *Reviving Ophelia*.

Mary Pipher and Ken Winston are leaders of "the group" in Nebraska.

A key phrase I found in Mary's manuscript (she let me have a peek) is "relationships always trump agenda." She goes on to say, "In fact, what I came to realize from my work with the coalition is that in individuals, families, communities, cultures, and even on Earth itself, nothing good and beautiful lasts unless it is grounded in loving, interconnected relationships." That sums up the group, and Mary. She lives as she says. I spoke with her at length after the meeting.

"The politics in this country are dead," she told me. "If we wait for politicians or international bodies or corporations to make changes, it will never happen. The world will be gone. The only way a democracy exists is when you create it every day. There's no way going to a voting booth gives people democracy. What gives people democracy is participating in a very engaged way in the decisions of the day. Democracy isn't just voting; it's making decisions locally about resources, about land and water."

When I asked her how the legislature could be so overwhelmingly in favor of the pipeline with so much local and national opposition to it, Mary related that TransCanada gave Nebraska legislators more than $800,000, over $600,000 of which came in the form of entertainment. "That's $42,000 in 'entertainment' for each legislator. What does that mean? What are they doing with all that money?" she said. "The people who actually have pure spirits and the common good in mind are almost never elected, and if they are, they don't last long."

The members of the group decided never to have a meeting where anybody left without something to do. "You deal with a lot of information," Mary said. "Upsetting information. It's very depressing if you have no way to act. I'm a worrier, but if I act, I let it go. I can go on being happy. We have all felt impotence, despair, sorrow, rage, and confusion. If each of us feels those emotions individually, we're whipped. But when we come together and realize we are all feeling this, there is an immediate catalytic energy, an immediate power. If someone has an idea, there are other people who want to hear it. If you *think* you have power, you have power, because you start acting like you have power. It's so simple to empower people; all you do is say, 'I think your idea is great, why don't you go for it.' It's like giving someone who's thirsty a glass of water."

Mary speaks seriously, with a mind that is always half a step ahead of the listener. But she pauses at the end of each verbal paragraph with a wide, cheery smile that lets you catch up and rest, briefly. She is always helping people through the heartache of witnessing the ecological destruction of Mother Earth and soothing her own spirit in the process. And she is a good judge of character. "We all know each other pretty well by now," she said. "We know strengths and weaknesses and who would do what. When we go to speak, my venue is the university, or the Unitarian Church. If it's speaking to ranchers, we want Randy Thompson to go. He looks like John Wayne—he can really talk to those people. One thing everybody knows about me is I always think a lot more people are going to show up at events than actually do. I'll think, 'This is such a great thing we have planned; five hundred people are going to show up!' and only fifty do.

"Churchill said something like, 'Success is going from failure to failure with no loss of optimism.' That's us. No matter what happens, we just keep working, we keep showing up. That's the most important thing: just keep working! The opposition never goes away, but they never wear us down. What keeps us from being worn down is having a really strong social network, a lot of validating each other. I have a vehicle here to hold my own anguish; something bigger than me. If all the pain I felt about the world stayed in me, it would be too much. This has been a way for me to put that pain into the care of a loving group of people. This has been a transcendent experience. The people who are most depressed are the people who know the most about the situation and don't do anything. Others are in denial. But what makes me a happy person is feeling I have a sense of agency and control. This group is one way I can do that."

I asked Mary what was likely to happen when the construction of the pipeline begins. "The place to take a stand is in your home, your environment, in your land. It's in Nigeria. It's in Kentucky. It's in Pennsylvania," she said. "It's everywhere. We feel part of people everywhere. We in Nebraska are at the hub of the universe; but so is everyone else. If this pipeline goes through, I think there will be an enormous amount of organized civil disobedience all the way from Alberta to Texas. And I think our group will be the organizers of that in our state. In most causes, you get the usual suspects to show up, but this one is going to be everyone. Not everyone is going to lie down in front of a digger, but we're going to have walks across the state, demonstrations, and dramatic events. We don't know now what we're going to do. It doesn't make sense to plan too far in advance. You're much better off having a tight team that works well together. Whenever something happens, they can come together quickly. I would certainly be willing to be arrested for this. I think thousands of people would be willing."

"Thousands?"

"Well, maybe hundreds ..."

I CAUGHT UP LATER WITH SEVERAL OTHERS IN THE GROUP. Aubrey Streit Krug is in her late twenties and a graduate student in English and Great Plains studies. She is currently learning the Omaha language, which very few people are still able to speak. She heard about the group through helping Mary with the *Green Boat* manuscript.

"I haven't been part of an organized group like this before," she told me. "It's exciting to meet people who share similar interests with me. I grew up in north-central Kansas, where my parents were dry land farmers [without irrigation]. We grow wheat, not corn. Maybe a little for silage, but it's not a commodity crop for us."

"Having just driven through Indiana, Illinois, the Dakotas, and eastern Nebraska," I noted, "I saw full sections, full square-mile fields of corn. I don't think we need any more corn."

"I'm actually quite interested in corn, as a crop for indigenous people, and I'm interested in the different varieties they grow," she said.

What Aubrey likes about the group is its informal nature. "It's social; it's organic. They're willing to shift interest and involvement depending on who's in the group, depending on what's exciting, what abilities they bring, and what they're willing to do. It's fun, and a lot less intimidating than joining a formal organization. I can miss a meeting, and that's okay. Mary invited me the first time—when somebody you already know invites you to come, that's a little different. I would have a hard time going if I didn't know anyone. There's spontaneity; you don't feel like there's a to-do list. And it's intergenerational. I find that very rewarding. Whenever you can get people in their twenties involved with really experienced people, that can be powerful."

"We didn't have that when we were in our twenties," I said. "The older generation couldn't understand us. 'What did we do wrong?' they wondered. 'You come from good families, we gave you everything you could want, why are you turning it down?' We would have benefited from more intergenerational contact."

When I asked her what motivated her interest in the pipeline issue, she said, "I feel a strong sense of responsibility and commitment to the Great Plains region. I don't mean that in the abstract. I have places and people I know and love that drive my academic work and my teaching. I want to learn and be useful, and taking action keeps you going."

ADAM HINTZ IS ALL ABOUT COMMUNITY AND ALTERNATIVE living. He was born in Lincoln and partly raised in a rural area nearby. He's in his thirties now, back in Lincoln, and owns the Meadowlark Café. He also hosts an environmentally oriented community radio program, *Earth to Lincoln*, on KZUM, where we taped a show on my travels and writings on the Keystone XL. After the show, I got back at him with a brief taping of my own. (I audiotaped all of these conversations.)

I began by asking Adam about his passion for living well on the earth.

"I have kids: A ten-year-old and a six-year-old, and I don't want to leave them in a world as horrible as the predictions are indicating, with climate change and so forth," he said. "Right now, our culture is barreling toward destruction. Anything I can do to stop that from happening I have to do as a father. I knew about this stuff before, but after I became a father, I became much more engaged. Now the XL pipeline has come along. Whatever I need to do about it, I will do."

"How have people here reacted to your activism?"

"I see more permaculture gardens here than before—a lot more ecological consciousness—but I still feel like an outsider here in Lincoln. This isn't Seattle or San Francisco. This is a *red state*. It's hard to be green in a red state. I love working with people like Mary Pipher and Jane Kleeb. I love them so much. I'm like a little Green Arrow working with Superman and Batman. To be able to learn from them and be part of what they are doing is a real honor. There's a real sense of community. It's another one of those subcultures within the larger movement. Our species is finite, like any other, but what we're working for is to give our species one more breath, one more inhale and exhale," he said.

Like many, Adam feels that society at large does not yet see what is happening to our world. He seeks comfort in the small society of friends he has in Lincoln, and he will continue to reach out with programs at the Meadowlark and KZUM, and by showing films about the Keystone XL in rural communities around Lincoln. Like Kermit the Frog, Adam doesn't always have an easy time of it, but he knows how to live simply, and to offer simple living to others.

Ken Winston is the group expert on the state legislature. By profession, he's a policy advocate for the Nebraska Sierra Club and spends a lot of time at the capitol building. His story is important, because it was Nebraska that alerted the nation to the perils of the Keystone XL pipeline, and people such as Ken who alerted Nebraska.

"It goes back to the Gulf of Mexico oil spill in April 2010," Ken said. "Nobody was paying much attention to the XL here, but all of a sudden people started thinking, 'There's oil in that pipeline, and it could spill out.' Senator Ken Haar, a real hero for us, asked me what questions to ask TransCanada in their application to the legislature. Then, later on that year, there were oil spills in Michigan and in the Yellowstone River in Montana. That got people thinking.

"Jim Pipher [Mary's husband] had written a piece about the pipeline in the paper, and I had heard that Mary was interested in the issue, so I joined this group. We decided to have a citizens' hearing on May 12, 2011, at the state capitol. Mary and Randy Thompson spoke. About a hundred people showed up. That same day, the legislature passed a bill requiring complete reclamation of the pipeline right-of-way. It wasn't a big deal, but it was a small victory. Then Senator Haar got twenty-one state senators to sign a letter to Secretary of State [Hillary] Clinton saying they did not want the pipeline running through the Sand Hills. In May, there was the leak in the Keystone 1 in North Dakota [the one near Bob Banderet's house], so things were happening. Haar called me up and asked if I liked the idea of a special session of the legislature for the purpose of moving the route out of the Sand Hills. The Sand Hills is something very valuable and very vulnerable. It's huge: the largest fresh water aquifer in the northern hemisphere, all fresh, clean water. I've been told it has as much water as Lake Erie, and it's very close to the surface. It's very important to Nebraskans, but nobody thought we would ever really get a special session. So we wrote letters to the editor, held art exhibits and rock concerts. Bold Nebraska organized a 'Shine the Light on [Governor Dave] Heineman' event at the governor's mansion. There were around seven hundred people with flashlights shining on the mansion urging him to call the special session. The Apple Pie brigade—several grandmotherly ladies in our group—brought cookies to the governor every week with a little message about calling the special session."

In March, a dozen Nebraskans went to Washington, D.C., and spoke to the Environmental Protection Agency and the State Department. Ken told me they asked for, and got, a supplemental environmental impact statement, which would require hearings and bring national attention to the issue. The Dalai Lama and several Nobel laureates wrote letters opposing the pipeline. Then, on the last day of August, Governor Heineman wrote a letter to President Obama opposing the pipeline because of its route through the Sand Hills. Twelve hundred people showed up at the State Department hearing in Lincoln. Ken wrote in his blog that day, "The last week has made me more proud than ever to be a Nebraskan."

"The Governor called the special session suddenly at a press conference in October," Ken said. "That was a victory for us. The session started November 1. There was a convergence of national and local events. Right as bills were being introduced here, the 'Hands Around the White House' event took place in Washington. It was crazy, but I went to DC—right in the middle of the special session—but I had to go. It was too important. I came back to Nebraska, and hearings started the next day. On our local television station, we heard Obama say, 'We can't sacrifice our water for a few jobs,' which was exactly our message. It was exciting to hear my words come out of Barack Obama's mouth! Then the State Department announced it would delay its decision for a year in order to evaluate the Sand Hills route.

"The special session hearings were overwhelmingly in favor of moving the pipeline out of the Sand Hills, but TransCanada was not giving an inch. During the debate, one of the leading senators asked me to come to his office. I thought he was going to chew me out for something I had said, but he wanted to make a deal. All the pipeline regulations we had come up with would not apply to the Keystone XL, but they would take the pipeline out of the Sand Hills. So we thought, all right, we'll take the deal. The downside is there was nothing official defining what constitutes the Sand Hills," Ken said.

"Then, during the regular session of the legislature this last January, they came up with a disappointing bill known as LB1161, which basically stated that the governor would approve the pipeline, there would be an environmental quality review, and TransCanada would have the right to use eminent domain. But this would apply *only to TransCanada*. It passed forty-four votes to five. This undid much of what we had accomplished in the special session. We're claiming now that it's unconstitutional because it is 'special legislation' that applies only to one legal entity.

"So, overall, we won round one of the Keystone XL battle, but they're coming back at us," he said.

When I asked Ken what was likely to happen if and when the pipeline is approved, he said, "Maybe we will get to a point where there is civil disobedience or worse, but right now I don't want to go there. I want to think in terms of how we stop it through the legal process. It's like kung fu: we take the opponent's mass, and we allow them to damage themselves. They're very prone to overreact—to using sledgehammers to kill flies. So I think we have to exhaust all our legal procedures. I'm hoping we get to the point where mining the tar sands is as unacceptable as apartheid in South Africa."

July 17, 2012 – Lincoln, NE
High 100°F – Low 71°F – Precipitation 0.0 inches

It's another scorcher here in Lincoln. It will be 100 degrees today, with a 30 percent chance of showers, but there is no other rain in the forecast for the next ten days, and temperatures will range from the nineties to 100.

A crop expert on the *PBS NewsHour* said last night that drought and heat stress during the tassel stage of the corn crop limits the number of grains pollinated, so there is no chance of recovery later in the season no matter what the weather does. But he said that Americans probably would not experience much difference in food prices, because so much of the retail cost of food is in packaging, transportation, and advertising. Only a small portion of what we spend at the grocery store ends up with the farmer, so when commodity prices go up, we don't feel much of a bump. Americans can afford to buy up the whole crop, which will push up food prices all over the world. The people who will feel the bump are the poor in other parts of the world. If half of a family's budget in Africa already pays for food, and food prices double ...

RANDY THOMPSON IS THE PUBLIC FACE OF BOLD NEBRASKA, literally. There are bigger-than-life fiberboard statues at stores and coffee shops around town showing Randy the Republican Rancher in his big cowboy hat with the words "I STAND WITH RANDY" printed boldly across the middle. He's tall and white-haired. Mary said he looks like John Wayne, but his manner is more Andy Griffith. He's not only the poster boy, he speaks well in public and to the farmers who most need to hear him. He's famous for phrases like "Water is clean or dirty; it's not red or blue." He has been in Washington several times for demonstrations at the White House and to testify at the State and Energy Departments. I caught up with Randy just outside of Lincoln.

Randy Thompson (left, with Ben Gotchall) is the public face of Bold Nebraska.

"This is the biggest issue that's come along in our state for a long time," he began. "TransCanada ran ads at a Huskers football game, and they got booed out. Their land agents act like, 'We're coming through whether you like it or not.' They gave us thirty days to accept an offer before threatening eminent domain procedures. This intimidated a lot of people. The way TransCanada has treated people is not the way we do business in Nebraska. They go to county boards and offer to build something at the fairgrounds or buy baseball uniforms or something—bribery, basically. A lot of people just accept their trinkets. We all know how it worked out for the Indians."

"How did your neighbors react when you took a stand?" I asked.

"I get a lot of support," Randy said. "At cattle auctions, a lot of guys come and say, 'Hang in there, we believe in what you're doing.' "

"I know your primary interest is landowner rights, but does the climate issue figure into what you do?"

"Climate change is a newfound awareness for me. Up until I got involved in this whole thing, I never thought too much about it. I always took care of the land, not overgraze, and that kind of thing. I didn't think too much about the environment in general, but when I saw what they're doing up there with the tar sands in Canada and all, I began to see the potential disasters we could have here with our rivers and streams. Yeah, it made me think about it."

"Does the word *environmentalist* still bother you?"

"No! I've come to admire them!" he said.

"We're not all communists, then."

"Yeah," he laughed. "Some people say we're being manipulated by Bold Nebraska—as if we couldn't think for ourselves." Randy is thankful for support from environmentalists. Property rights remain his key motivator on the pipeline issue, but when he found out about the tar sands, he suddenly saw a much bigger picture.

"What about what we're experiencing here this summer?" I asked. "We've had droughts before, but this one is bigger in scope than what has happened before."

"I personally think climate change is real."

"What do you think things will be like around here when construction begins?"

"It will be very interesting," Randy laughed again. "There are a lot of folks pretty upset about this. They may make a physical protest. I'm not talking shooting people, but a protest. A bunch of us went to D.C. in November for the 'Hands Around the White House.' John Harter [from South Dakota] was there, and we got to talking, how it affects you. I mean, I was getting up at three o'clock in the morning, couldn't sleep. And he was going through the same thing. I talked to Paul Mathews, too. Same thing happened to him. I mean, it wears on you. I don't know, we might have 50,000 people at the border in Montana. I'd make a stand."

"I wasn't in D.C. in November, but I went in August."

"Did you get arrested?"

"Yeah," I admitted.

"Good for you."

"And I don't even live here."

"Well, this is a whole lot bigger than just Nebraska."

"But you guys are on the front line," I said. "You're the ones who brought it to national attention. You're organized.

"What about civil disobedience?" I went on. "If it was done well, if people were polite, gentle. How would it feel to you if people came here from the outside?"

"I'm not in favor of that," he said. "I don't think the people of Nebraska would accept that; we can fight our own fight. I don't want you to think we don't appreciate the support, but people in Nebraska are going to listen to other Nebraskans. ... But, you know," he paused, with a laugh and a big grin, "when it comes down to the point of construction we might need all the help we can get! ... Come on down, Sam!" He laughed out loud.

As I got up to leave and shake his hand, the last thing he said was, "You know, this really grates my soul."

JOE MOLLER WORE A BASEBALL HAT AND JEANS, AND SPOKE IN a slow, deep, thoughtful voice. He owns eighty-eight acres west of Lincoln that looked to TransCanada like a nice place to put a pipeline. The land came from his parents, and it will pass on to his kids and grandkids. He leases out part of it for grazing, but it's mostly a weekend getaway for the family. It has a beautiful clear lake for fishing and swimming and has become the center of extended family life. Joe's daughter, Kim, is passionate about saving it from the environmental uncertainties of the Keystone XL pipeline. Both are members of Bold Nebraska.

Joe worked for Northern Natural Gas, a regional company that got bought out by Enron. "Lost two-thirds of my pension," he quipped. "But that's another story.

"I know about pipelines," Joe said. "They leak. They might not leak for several years, but they leak. Tar sands oil is corrosive and abrasive. It's not like natural gas. They get a lot of the sand out before it goes in the line, but they don't get it all. That silicate will scrape the inside of the pipe; it's abrasive and corrosive, too. Up in Canada, they have these gigantic shovels that mine the tar sands, and it's so corrosive they have to replace the steel teeth on the shovels every day. The pipeline is made out of steel less than half an inch thick. And they put all kinds of chemicals like benzene in the tar to make it runny so it will flow."

Joe told me that the water table is about four or five feet below the surface in the Sand Hills, and the bottom of the trench they dig is a minimum of seven feet, so the pipe will not be over the aquifer or above the aquifer; it will be *in* the aquifer. It will be sitting in the water. They can detect a leak of 1 or 2 percent, but that amounts to tens of thousands of gallons of tar sand. "Generally, where you have a leak is not on the weld, but right next to it, where the metal has been heated and weakened," he said.

And Joe is looking ahead to the day when the pipeline is no longer in use. "Once they determine the life cycle of the pipeline is over, the landowner becomes responsible for any leaks! When they stop pumping, there's still going to be residual oil in low spots. There's no way to get it out. So when the metal eventually corrodes away, the landowner is stuck with the liability and for cleanup costs. That might be fifty years from now, but that's my family."

Kim and Joe Moller are passionate about saving their land outside Lincoln.

"I'm the future landowner," Kim piped in. "This is something my grandparents worked hard for so we could enjoy it. We go out there to relax, and there's nothing relaxing about this pipeline."

"The pipeline was not part of the original vision," I joked.

"No, not at all!" she said with a giggle.

Kim brought out a four-inch-thick notebook collection of all the letters she had written to the legislature, the Public Service Commission, the Nebraska Department of Environmental Quality, the State Department, and the president. She had three other binders back home. "Senator Avery wrote back to me that all it takes is one teaspoon of tar sands to pollute a whole swimming pool of water beyond drinking water standards."

"That's my name, too," I said, "but I don't think he's any relation. Sounds like he is on your side."

"He was, but then when LB1161 went through, he voted for it."

When I asked her what motivated her to work so hard against the pipeline, she said, "Mainly, it's my family. I want to protect our land and our water. Some people don't understand, but I've gotten other people on our side just by telling them about the consequences. We know about this sort of thing, because we're first responders. I'm a firefighter, with my dad. TransCanada doesn't have to release any hazardous materials data sheets on what they have going through the pipeline."

"I don't know how they get away with that!" I blurted.

"We don't know if we need hazmat suits," Joe added. "We don't know whether to go in or not. Are there dangerous fumes? Do we need oxygen tanks? They won't tell us. They call it 'proprietary information.' "

"Do your neighbors support your stand?"

"Some do, some don't; most of them are neutral," Joe said.

"But the neutral ones sign on," I stated.

"That's right. I've got a neighbor to the south of us who's all in favor of it."

"I'd like to talk to him."

"Curt Carlson. I'll give you his number."

"Great. I'll find him. But I want to know if this whole thing has changed your understanding of things like the environment, or the climate."

"It's definitely making us more aware of the environment," Joe responded. "This gives you a wake-up call. I was like 95 percent of the rest of the country. I thought climate goes in cycles, but I'm thinking now that it's changing."

"We're still going to have cycles," I said. "But they will be hotter than before; the storms will be stronger, the droughts drier. Whatever the atmosphere was going to do anyway is just going to be worse."

"I would love to have a wind farm on that land," he said.

"What will happen when construction starts? Do you think people will physically resist it?"

"Yup, I do. It could get ugly, really ugly." There was a tense pause. "I don't want to see it come to that," Joe said.

Kim had been on her cell phone, but perked up when we mentioned resistance: "I want to put that on my résumé! I fight for my family!"

I noticed a mild generation gap developing across the table. "If it gets violent," Joe said, "then everybody's going to think we're a bunch of tree-hugger nutcases."

"Usually when people call us tree huggers," Kim said, laughing, "I tell them, yeah, maybe if I've had a few too many drinks ... We went to D.C. and surrounded the White House. If it weren't for the August protests there, I don't know if we'd be talking here now."

"We met a lot of people there who looked like professional protesters," Joe muttered, "but there were a lot of genuine people there, too."

"So what's your status now? Have you signed anything?" I asked as I stood up to leave. Joe gave me Curt's phone number, and Kim took my e-mail address to send me more information.

"Have we? No. Will we? No."

I FOUND CURT CARLSON SHORTLY AFTERWARD.

"There are people here who want TransCanada to move the pipeline onto their land," he was telling me. "'Here! Put it on our side of the fencerow! We'll take the money!' It's just the Bold Nebraska group of individuals that are screaming the loudest. Jane Kleeb's husband got pounded in a senate race, and now they got to go around hollering about the pipeline. That's all it is. And Obama doesn't like oil. He doesn't like coal. He's about to shut down the coal industry."

"Yeah, I know, I'm from a coal state," I said.

"Yeah, I saw your bumper sticker." (*Clean Coal is like Dry Water.*) Curt had a ready answer for every question. There was a little swagger in his manner when he was explaining himself, and he tended to start the answer before the question was fully asked. He often interrupted toward the end of my sentences with an up-volume "Well ..." or "What people don't know is ..." It might have bothered me, but he was always kind and polite, and never hostile. He really wanted to have his say. His readiness with answers reminded me of some of the anti-pipers I had run across earlier. Curt turned out to be a really nice guy.

"You see that corn over there?" he asked. "That's food. You know what it needs? Carbon. What are we going to eat? Most people don't understand ... that fire in Colorado two weeks ago created more greenhouse gases than we've made in the last forty years. God creates more carbon than you, me, and the neighbor create with coal and oil ... Even if we burn all the coal and oil, there are more trees in the continental U.S. than there were in 1650, or whenever. We've got more green now than we did then.

"The pipeline will bring jobs," he continued. "We're running refineries at 70 percent capacity now. If we run another 30 percent, that means more jobs and more refined product. And it's national security. If they shut down the Straits of Hormuz, we've got Canadian oil."

"Yeah. It's not ours, but it's from a friendly neighbor."

"And it wouldn't be real hard for us to make it ours, if you know what I mean. And I don't mean to say we would do that. The other thing is, do you want TransCanada building a pipeline to the East Coast? We'd rather have it through here."

I thought it was time to try my usual climate pitch. "When we first started burning coal, gas, and petroleum 150 years ago, there was about 280 ppm carbon dioxide. Now it's around 393. That's 40 percent more carbon in the whole planetary system than we had before. That can't help but change the way the atmosphere works," I said.

"Yeah, but take Mount St. Helens," he answered. "That put more carbon,

"The pipeline will bring jobs," says Curt Carlson.

ash, and soot into the air than since the beginning of the industrial age back in the 1880s, 1890s. And these people who talk about carbon: they get in their cars and drive around, they get in their private jets ..."

"There's a lot of hypocrisy out there."

"Yeah, and I don't think carbon's the problem. The pipeline doesn't have anything to do with the amount of carbon. Now, look at that pivot [irrigator] over there. If you want to eat, we've got to generate the energy for that."

"You can generate electricity without fossil fuel," I responded. "I'm not just talking off the top of my head. I'm a solar installer."

"Not at night, you can't."

"Right. You'd need an energy storage system, or a backup."

"Come over here; look at this," Curt said. He led me to a shed out back and pointed to a charge controller and inverter mounted on the inside wall (the charge controller regulates the voltage from the panels to the batteries, and the inverter turns direct current into alternating current.). "Oh, I know all about solar. This is a pure sine-wave inverter, only 600 watts. I'm just playing with it here until I find what I want to do later." Back outside, he showed me a south-facing roof with the right pitch for a solar installation. "I'll get around to it. But right now it's not economically here."

"It could be if we invested in it. Pipelines and coal mines don't make jobs; investment makes jobs. If we invested in renewable energy, we would have the jobs and the energy without the pollution and climate change."

"Yeah, but Obama tried that already with Solyndra. He dumped billions of our dollars into ..."

"That was because the federal government made a stupid investment," I argued. "That's not because the whole industry is bad."

"Well, I'll tell you what, when it's there, it will happen, and the government won't have anything to do with it," he said.

"I would hope so. But the government is subsidizing fossil fuels. Over a hundred billion dollars in the next ten years."

"They lost that much with Solyndra."

It was half a billion, but I let that one pass and moved on. "If they didn't subsidize fossil fuels at all, I could compete with coal and natural gas, even petroleum. I could install systems that would be cheaper in the long run."

"But the price of food would go way up ... 'til somebody figures out how to make hydrogen cheap."

It was time to change the subject. "Do you have any concern about local environmental effects of the pipeline? Are you worried about spills?" I asked.

"No," he answered quickly. "There are already two hundred thousand miles of pipelines and gas lines crisscrossing all over the aquifer. One more little pipeline isn't gonna make any difference. They're making a big deal out of nothing."

"Has TransCanada treated you right?"

"Yes. They're paying a fair price for easements. I met the president of TransCanada. Everybody's been nice. I went out there one time, and there was a crew digging around in one of my springs. I said, 'You can come through here, but you're not going through my spring.' "

"Did they listen?"

"Uh huh. One of the engineers told me they were going to need forty people right off the bat. I might just ..."

"For construction?"

"No, for maintenance. They need people to drive the line, manage the station, welders—the whole nine yards."

"What do you think about this weather we're having now? Is this just a normal trend?" We were still standing outside, and he was sweating profusely.

"It happened in 1988," he said.

"I remember it well."

"We've had ups and downs. Three years ago, we had so much snow we didn't know where to put it. What people don't realize is that God is in control; they're not. God created this place. People are trying to humanize this world, and if they can say man is in control, then there's no God, and they can justify their actions. Until they get their arms around God, they will be fighting everything He created. He created us. We didn't create this," he said, and paused as he looked up into the deep blue Nebraska sky, raising both arms above his shoulders. I thought he had forgotten me for a moment, but when he looked back my way, his face said that I was in his world. "God didn't put us here so we could screw it up ... too bad ... but we might." Then his expression changed.

"They're atheists, and through this global warming and everything, they gotta say ... they got to strangle out *God* so they can justify their homosexuality. They're a bunch of guilty little suckers. Nine-eleven was a shot across the bow, just like what He did to the people of Israel before He sent them into captivity. God was warning us to turn from our evil ways."

"Do you think we could have some ways that we're not seeing? Maybe not evil, but things we're doing that are destructive of the creation God has given us?" I asked.

"He gave us dominion over the world."

"But dominion has responsibility. It doesn't mean you can do whatever you want. I feel like we are in God's garden, and we have to take care of it."

"That's why I'm satisfied with what TransCanada did. They had bug-ologists, geologists ... I'm not really all that miffed if we lose another species—I hunt them and I eat them. Let me show you something over here ..." We walked over to the barn to look at his horses. "Got some thoroughbred in 'em," he said, patting the nose of his favorite.

I admired the horses, and we talked about the hay crop. As we left the barn and headed back toward my car, he said, "It's all about control. They want to control food, control oil. If they control food, they control you."

"Who's they?"

"Part of it's the UN, part the Democratic Party. It's an international deal."

As I opened the car door, Curt picked up a small gray kitten and stooped to look behind the tires. "There are two of them, twins," he said. "I just saw the other one around here a minute ago. They're not smart enough yet to stay away from cars."

I closed the door without getting in. "Let me see that little thing. I'm a cat person." Curt grinned and handed her to me. She was strong and bright-eyed, and soon she relaxed in my arms. "This cat's well taken care of," I said. She was purring already.

"Yeah, I really like 'em. Let me check under the car again."

"I'll go slow," I said, handing back the kitten. "Thanks for your time."

"Thank you. Have a safe trip."

July 20, 2012 – Grand Island, NE
High 98°F – Low 70°F – Precipitation 0.0 inches

I'm back on the road, west of Lincoln: fewer trees, more prairie. And more irrigation from the aquifer—the ban on watering applies only to surface water. Days like this bring into focus the vital importance of Nebraska's Ogallala Aquifer.

I'm camping for the most part on this trip, if you can call it camping. No tent, no RV. I just fold down the seat and curl up in the back of my old Volvo wagon. I'm here in Grand Island for two nights. This car's a real camping machine. I have all I need: mat, sleeping bag, cooler, food, beer. I even have a little inverter and power strip hooked up to the battery with all kinds of WiFis, cell phones, computers, and camera batteries plugged into it. I've driven this car through 280,000 miles of interstates, gravel roads, puddles, and potholes (eleven times around the planet), and it still drives like a new Lexus. It has a few dents here and there, but no broken handles and no holes in the upholstery. I just hope it falls apart in time to get my plug-in hybrid. I'd hate to miss the solar age because my Volvo wouldn't bow out gracefully.

THE PARK ATTENDANT CAME BY THIS MORNING TO WATER THE grass near my campsite. "Any relief in the weather today?" I asked.

"Nope. Gonna be around a hundred again. No rain till Monday or Tuesday, and that's only 20 percent. This drought's everywhere now, all over the country. I haven't seen anything like it for thirty-seven years."

"Do you think the climate's changing?" I ask this of just about everybody I meet, to pass the time and to see what's floating through the public mind.

"Yup, I do. Ever since we started flying up around the moon, and so on." He waved a hand in the air over his head. "We messed up the atmosphere up there."

IF RANDY IS THE FACE, JANE KLEEB IS THE MUSCLE OF BOLD Nebraska. She's been in Nebraska less than five years, which counts against you in these parts.

"The Republicans—well not Republicans, but the Republican *Party*—doesn't like me here," she started. "There are blogs against me; they call me 'Insane Jane.' I had no environmental background, no energy background. I founded Bold Nebraska in March 2010, focusing on health care reform."

In May 2010, a friend of hers, Duane Kavorka from the Nebraska Wildlife Fund (who was at the meeting at Mary Pipher's) told her about a tar sands pipeline coming across the Sand Hills and asked if she wanted to help organize against it. There would be a lot of local and national attention around it, and a hearing was coming up in York, Nebraska, the following month.

"I was thinking, 'Who's going to go to a State Department hearing on a pipeline nobody had heard of?'" she said. "But 150 people showed up, and everyone who came up to the microphone was against it: farmers, ranchers, moms, grandmoms. That's where I met Ben [Goschall, a current co-worker] and several others. We kind of huddled after the meeting and said, 'We've got to do something about this; this is crazy.' We didn't know what tar sands were. We didn't even know there was already a tar sands pipeline in our state [Keystone 1]. When we started talking to landowners, we found out they didn't know what was going through the pipe on their land. Many of them thought it was gas. Some of the landowners along Keystone 1 had concerns a few years ago about granting easements, but they had no one to turn to. They felt alone and powerless."

"Was the initial opposition just for landowner's rights? Or were there any environmental concerns?" I asked.

"The initial concern was the aquifer," Jane said. "That's an environmental issue, of course, but folks don't immediately talk about it that way. They talk of protecting the land and water that is their family heritage. They don't think of it as 'environmental.' They know it's environmental, but they don't use that word."

"But the primary fear is of a leakage?"

"Yes. And of the construction process itself."

"So maybe we just need a new word. Sometimes I feel the label *environmentalist* is getting in our way."

"Yeah. It has a different meaning here," Jane said. "Nebraskans consider themselves *stewards of the land*, or *conservationists*."

"But do you feel your environmental awareness has opened up as a result of this issue?" I asked.

"It has. I knew about global warming before—I was kind of your typical progressive—but it really means something to me now. There's a story about a rancher who was bullied by some politicians when he came to testify in the legislature. They were asking him things like, 'Who's paying you to be here?' He called me up on the way home and said, 'Ya know, I just pulled over and hugged a tree!' So I felt myself doing that, too. Instead of running away from being called a tree hugger, I'm proud of protecting our environment and our community. I mean, good Lord! If you don't protect the environment, there's nothing to pass on to our kids, and our kids can't grow up healthy."

Jane Kleeb is the muscle behind Bold Nebraska.

When I asked her about the status of the approval process, Jane said that the route would be finalized and easements acquired by early 2013, which meant that if Romney won, construction would be likely soon thereafter. "What if Obama wins?" I asked, expecting the usual cynical answer.

"I think he will say tar sands is environmentally safe, but that the pipeline is not in our national interest because of all the new oil discoveries in North Dakota and elsewhere. He's likely to say we have our own oil to move—they won't let our oil in the XL," she said.

"Huh ... I have been thinking Obama would approve it with the new route."

"Another interesting thing about this issue," Jane said, "is you see folks talking about connection to Native Americans and tribal communities, how we took their land away from them, and now we're seeing it happen to us. Ranchers seem to have deep connections to American Indians."

"Yeah, I was thinking about that today," I said. "Something Randy said about the American Indian experience made me think how what we did to them was not unique—genocide is a regular feature of human history. What's unique is that now we have reached a level of maturity as a society where we can look back and see how wrong what we did was. Now we're beginning to identify with indigenous people—they're us! We feel great pain for what they went through. That makes me proud to be American."

"Yup. Me too," she said.

I asked Jane my usual question about what was likely to happen here if and when construction begins.

"There's no doubt in my mind that folks will lie in front of bulldozers, just like in the rain forests. I think you'll have the same images. Instead of native people walking in front of the bulldozers, it will be ranchers and farmers, with the urban moms and grandmoms standing right next to them. There's no doubt in my mind."

"What about outsiders coming in? People like me? What would you think of that?"

"I don't see you as outsiders. I think the landowners and ranchers would feel the same way."

"I've gotten a variety of answers on that one. I'm an outsider to Nebraska, but I'm not an outsider to the atmosphere," I said.

I'M LEAVING NEBRASKA TOMORROW TO SEE WHAT PEOPLE ARE up to in Oklahoma and Texas. This trip is different; instead of the usual east-west route through the Great Plains, I'm going north-south. I'm here to be here, not just driving through on my way somewhere else. And I'm beginning to feel at home, a little less like an outsider.

Oklahoma

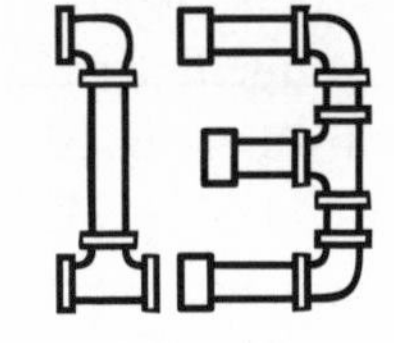

July 21, 2012 – Edmond, OK
High 105°F – Low 75°F – Precipitation 0.0 inches

It's drier down this way. The wheat crop is in, and it has probably done well. A cutting or two of pre-drought hay has been raked and pressed into huge plastic-covered bales, scattered like giant Tootsie Rolls across the gently heaving prairie. The unirrigated corn at the edges of fields and corners where pivot irrigation systems don't reach looks bad—real bad. I saw some totally browned-out fields that looked like corn in a Halloween decoration. The price of corn goes up every day. Randy Thompson in Nebraska told me the other day he sold some partially irrigated corn for $7.45 a bushel, and Curt Carlson said he expects the price to go up to $8, maybe $9. That's good news for people who have something to sell.

I VISITED TODAY WITH HARLAN HENTGES, A LAWYER FOR THE Center for Energy Matters in Edmond, just north of Oklahoma City. The center was originally organized around a drive to stop a coal-fired power plant at Shady Point on the Arkansas border. It has a five-member board and two staff: Harlan and Rosemary Crawford, whom I met later in the day. The organization is grant-funded, and its mission is purely legal assistance; it is not a membership organization. So far it has been involved in stopping three coal-fired plants in Oklahoma.

Harlan is a round-faced, robust, big-hearted man with a giant Oklahoma smile. He is progressive and open-minded, with a strong conservative streak. His background is in farming and ranching. He is especially concerned with the state legislature's grant of eminent domain to TransCanada. "The Keystone XL is not a public use facility, and eminent domain in this case is unconstitutional," he said. I asked him who he contacts in his everyday work.

"Whenever it's eminent domain, I team up with radical right-wing property groups; whenever it's the environment, I team up with radical left-wing environmental groups. As near as I can tell, they have the same set of values. The distinctions between those two, and the knee-jerk reactions in opposition on an ideological basis, I think, is all divisive and fictional. I don't think that division has any merit. If you look at what's going on and say, 'Is this good? Is it just?' And if it's not, why are we doing it?"

"Environmentalists want to protect the environment generally," Harlan continued, "and landowners want to protect the environment specifically. They want to protect a particular piece of the environment, and I think that's how you do it. If you can't get those people to protect their land, then your attempts to protect the environment generally will be thwarted. If you want the government to protect the environment ... well, that ain't going to happen, because the government is largely influenced by corporate interests. Corporate interests are *by law* the financial interests of their shareholders."

"But something like the climate, the atmosphere," I interjected, "cannot be specific. It can only be general. Do the climate effects of the XL pipeline ever enter into your conversations with people?"

"As a practical matter, I never talk about climate change, unless I'm doing it to create conflict," he said. "I'm somebody who's relatively concerned and interested in the impact of carbon on the climate, but I don't talk to anyone about it because there's nothing to be gained."

Harlan did not expect much opposition once construction begins. "Not in Oklahoma. This state is the reddest of the red." There was a long pause. Harlan seemed apologetic for the lack of environmental awareness in his state. The *environment*, as a state of mind, simply does not exist here. The issue in Oklahoma, to the extent that it exists at all, is the heavy-handed approach of TransCanada in its dealings with landowners.

"They act like they're from a British colony," Harlan said. "They act like the power resides in the sovereign."

"George III."

"Yeah."

"Didn't we win that war?"

"Yeah," he laughed. "I thought we settled that! But you let a dispute like this go on long enough and you end up fighting a war over it. Yeah," he paused again, rethinking, "You are going to have people opposing it. When somebody comes here and tries to take away their rights, you'll have left-leaning people who are well trained in direct action and you'll have some right-wing people

"As a practical matter, I never talk about climate change," says Harlan Hentges.

who aren't. They're gonna have guns. If you're preparing to do this stuff, you ought to understand this possibility."

I thought for a minute about people defending their land with weapons. It would show resolve, but would do little to change public opinion as a whole. "I would say to them," I said out loud to Harlan, " 'Good for you for standing up, now let's figure out how we can do this effectively.' "

"That's the ultimate failure," he said. "As a lawyer, I have to say that lawyers are the alternative to guns and soldiers. Even direct action or nonviolent protest is a failure of the lawyers."

"I think you're absolutely right that the legal system is the way we avoid confrontation, yet often it's through confrontation that consciousness changes, whether that's violent or nonviolent," I said. "In the American Revolution, which was violent, we became Americans. We became something different from what we thought we were. And in the civil rights movement, which was accomplished nonviolently but through direct actions, we changed what America was. I remember thinking at first that civil rights meant we should be nice to 'Negroes,' but then, over time, as I watched what people were doing, I started thinking, 'I'm the *same thing* as those people sitting at the lunch counter. That's *me* sitting there!' "

"Exactly ..." Harlan was grinning sheepishly. "I think I got goosebumps there ... That's exactly right."

"Sometimes confrontation, even outside of the law—if it's done properly, and with wisdom as to how to engage—can create a different sense of who we are."

"I look forward to that, and I think it will happen. We've got a real poor concept of who we are," Harlan said.

THAT AFTERNOON I DROVE EAST AND MET ROSEMARY CRAWFORD at Shawnee. This part of Oklahoma is Indian country, home to Cherokee, Shawnee, Choctaw, Chickasaw, and Kickapoo. Rosemary has dark, straight hair but a light complexion, and I did not at first see Indian in her face. Her mother is Cherokee and Choctaw, while her father is Dutch. She was raised in this area but spent many years in California. She returned eight years ago for family reasons but misses California. "People here don't think outside the box," she said. Rosemary went to work for Harlan in 2008 on the power plant issues and later began working on the Keystone XL pipeline. "I still try to help him when I can, but I've been in respite for a year now because I have not been pleased with the negative side of the fight."

"The negative side of the pipeline or the opposition to the pipeline?"

"Both sides. I was feeling internally as if I just wanted to escape. I was doing everything at the center: talking to landowners, talking to media, driving around. There wasn't anyone else to delegate to."

"Up in Nebraska, people are bouncing off each other a lot," I said. "Sort of like my group in Louisville, they have a strong sense of community. They party ..."

"And I was never able to form a community," she said. "I'm kind of out of my element here, but I believe I was placed back here because there is a need. There's an awakening that needs to take place."

We climbed into her SUV and headed for the Red Mound Ranch, where the XL will cross sometime this fall. Jack Landrum and his wife own nine hundred acres on either side of Interstate 40 north of Wewoka in the Seminole Nation. Jack is a former oil worker. After retirement, the Landrums bought the land to start a cattle-breeding business. The pipeline will run diagonally across their land and within a hundred yards of their house. TransCanada will put a large equipment yard in the front yard, by the road. Jack will not be allowed to drive his own equipment across the easement, which will compromise the utility of the land to the point where he will have to move the operation to keep it going at all. He is too old to consider that possibility, so he will simply close it down. When I asked if we could talk to him, Rosemary said that he was under a gag order. Apparently, as part of the settlement agreement with TransCanada, he was not allowed to speak to anyone about the company or the pipeline. He made a statement on TV last year opposing the XL and was reprimanded by company representatives.

"Their dreams are gone; their spirit has been taken," Rosemary said sadly.

As she recalled Jack's reaction when he found out they were going to take his land, Rosemary became more emphatic. "He's in his eighties," she said, "and he's in tears because a private company from a foreign nation has come in and taken away his ability to have free speech in the *United States of America.* This is what we believed in, this is what we were raised on, and I have seen people shaken with this to the core. This goes against *who he is* to the point where he's not going to choose to be here anymore. He was an oil worker; he believed in oil companies. He believed in *progress.*

"Jack showed his emotion to me one time when we were standing on a little hill outside his home," she continued. "He was looking at the cattle. He was talking about his life, about being a steward of the land. I was telling him that this was what it meant to be an environmentalist—loving the little piece of joy and heaven that you have here. 'You're a steward; I'm an environmentalist,' I told him. 'It's the same.' "

We drove through the Seminole nation and into the dying town of Wewoka. Buildings were boarded up on either side of the main street. Rosemary spoke of her childhood.

"I was raised here. My ancestors walked the Trail of Tears." Her grandfather was a Choctaw and grandmother a Cherokee. Her great grandmother, Amanda Glass, walked the Trail of Tears from Tennessee as a child and arrived in Oklahoma at the age of seven. Not all of the family survived.

"This is the courthouse," she said. "This great big tree you see here is called 'the Hanging Tree.' One time when I was little, I walked by here with my grandfather and he kind of shuffled me along past it. Years before, a local judge had ordered a lot of Indians hanged right here. 'Now don't ever tell anyone you're an Indian, 'cause they might hang you,' he said. I was born with fair skin. His side of the family was dark-skinned, but my grandmother was Cherokee, so her skin was lighter and her eyes were lighter. I was born with blue eyes. They paraded me around this town like I was a prize."

"You were able to pass."

"I was able to pass."

On the way back to Shawnee, Rosemary was anxious to say more. She wanted to get her story out.

"We are moving toward a new paradigm," she began.

"Did you know that word was in the title of my book?"

"No ... well, I remember it now in the proposal, but I wasn't thinking of it." She was getting pumped up, so I backed off and let her have the floor. "And I believe the shift in our consciousness as a planet is moving us truly into the Age of Aquarius and into a feminine ... more masculine/feminine balance. There has to be a major shock in some form for the masses to get it, and this is a part of it."

"You mean the pipeline?"

"Not just this pipeline, I mean the extraction of fossil fuels as a whole in such mass—tar sands, oil, mountaintop removal, fracking—and that the people who are benefiting financially are in denial. They want to continue to be the rulers of the world. They think they are in control, but nobody's in control. Our existence on the planet is not going to be any longer controlled by those people who have manifested the largest amount of money. At the end of the day, how much money you have means nothing. To come back into balance on the planet, where this Earth is not *screaming* at us, and we're not living in a place of constant duality, this shift has to occur, and occur now. It is by design; none of it is by accident. My personal work is to raise the conscious level of the masses—the people who are not of a high level of wealth and who suffer the consequences of the extraction of fossil fuels, the burning of the coal, the fracking—the people who are suffering and are considered of no value. They are expendable, they're on a spreadsheet somewhere, and their lives don't count. I believe it's those people that we awaken and bring into consciousness that they are as important—as a spirit and a soul—as the man with a million dollars who thinks he's better than they are. By doing so, by waking up and equalizing the male-female energy, we will bring our planet to the place it needs to be. I may be naïve, and I may be the only person who believes this, but that's what I genuinely believe, and that's what I'm here to do. I believe I was sent back here to do this work."

"Opposing the pipeline?" I asked. "That's essentially negative, it's opposing what somebody else wants to do. It's not really creating any new consciousness. So how do you see a new consciousness, a new paradigm developing from this struggle against the Keystone XL pipeline?"

"I don't," she answered. "That's the reason why I was so uncomfortable."

"Oh."

"I was part of the negative energy. I was attracting negative energy by focusing on the negative side of the problem. By doing so, I was *increasing* the problem. I became in turmoil. I had this constant frustration and anxiety: all those low-vibration energies just consuming me every day. I was focusing on what I didn't want, which is what I was *attracting*. The universal law of attraction says what you focus on is what you get! So I'm attracting the pipeline because I'm out verbally speaking my truth, telling the universe these 'n' words and the universe doesn't know these 'n' words, so it's going to attract what I'm saying I *don't* want. The universe is going to read that I *do* want it."

"What are 'n' words?" I asked.

"No, not, don't, can't ... all of that. The universe doesn't know what they mean. So I needed to shift what I was doing. I felt internally the tipping point has already been crossed—in the 1980s—it's not just tipping now; it's dumping, over the edge.

"My three grandchildren," Rosemary continued, "as long as they have a body, are going to need food to eat, water to drink, air to breathe, and right now that is not possible if we keep going down this road. My great grandparents, when they got to Oklahoma, had no money; they had each other, and they had air and water and land. But my grandchildren won't have that. The basics of survival on the planet are being destroyed if we don't make some conscious decisions to change that. As we have a body, we have to feed it, nurture it. Our minds, our bodies, our souls: all are one, and we are all collectively as one."

"Let me go on from there." I interjected. "I follow what you're saying, but in your attempt to manifest your awareness of what's happening to the earth, you're tapping into some negative energy."

"Yes. It's huge."

"I'm wondering if that has to do with your feeling like you're not doing this in community."

"Yes, that's exactly right," she said.

"You're feeling it individually. You've got your own shield out there, protecting yourself individually, but you don't have people on either side of you with shields."

"So I have been doing visualizations—visualizations that the pipeline is not built, that it is stopped, because we have enough Earth energy that shifts to a place where digging a hole in the ground and laying a pipeline is no longer profitable, or conceivable. And I do visualization of going up above the planet and backing myself out from the earth, and seeing it spinning clearly with clean water and clean air, and vegetation. I visualize new worlds and new growth and a new spring, and the people who are still living on the planet understand the needs not only of themselves, but of the earth, and that the seed—as you put it—continues to grow. Everything goes back to being whole, healthy, and complete. How many thousands of years it takes for her to revert back to the way she was, I have no idea. The duality of male and female energy will cease."

"The yin and the yang."

"The yin and the yang."

Texas

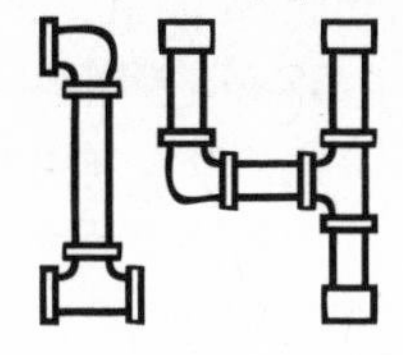

July 22, 2012 – Winnsboro, TX
High 98°F – Low 73°F – Precipitation 0.0 inches

I woke early this morning at Murray Lake State Park in southern Oklahoma. After a short jog through the campground and quick dip in the lake, I was ready to move on. I drove south passing small trees, low hills, and browning pastures. There are virtually no row crops in this part of the state. The cattle are huddling in patches of shade below cedars and post oaks, refusing to fatten themselves in the hot open fields. There are not many of them, and there will not be much for them to eat this winter. The Agriculture Department reports that there are fewer cattle on American soil now than at any time since records have been kept. Beef will be cheap this fall as ranchers cull their herds, but prices are likely to skyrocket next year.

The small trees and rolling hills made me feel I was arriving back east, but the land turned flat and the trees thinned as I crossed the border into Texas. A few scrubby oaks and pines persisted, but I was back on the prairie. Turning east from Denton and traveling across the north Texas plain, field after field of corn was totally browned out. Mile after mile, there was not one healthy stalk, and many fields were already plowed under. It was the worst crop devastation I have ever seen. Some north Texas mortgages will not be paid this fall.

I MET DAVID DANIEL, HIS WIFE, CLARA, AND FOUR-YEAR-OLD daughter, Naida, at their twenty-acre homestead just south of Winnsboro. The XL will cut their land in half. The southern leg of the pipeline (from Cushing, Oklahoma, to the Gulf of Mexico) has become a separate project from the northern leg through Montana, South Dakota, and Nebraska. It does not cross a U.S. boundary, and therefore did not require State Department approval. It was set to go, while the northern leg waited for presidential permission (and a presidential election). The middle section of the southern leg—through Winnsboro, where I was now—needed final approval by the Army Corps of Engineers, but TransCanada announced that construction would begin in the sections through Oklahoma and south Texas on August 1. That was in ten days.

From the house, David and I walked down the path to a small camping area halfway down the hill. Two or three tents were pitched under the trees, but no one was around. At the bottom of the hill a small, slow-moving stream passed between one-hundred-foot oak and hickory trees. Crows were cawing in the distant hills and squirrels were cackling in the trees nearby. The air was still and hot. This was where the Keystone XL pipeline and millions of barrels of tar sand would cross David's land. "That's the tree village," he said, pointing to a rectangular platform high in the branches overhead. "This one's mine. There are five others farther down the stream. We plan to be up there for a long time: there's plenty of food and water there already. I'll take you up when I get the chance, but I have to go now."

"Did you build all this?" I asked.

"With help. Mainly from Ron."

Back at the campsite, Ron Seifert appeared at the kitchen tent. Tall and lean, with a short beard, dark hair, and sharp, handsome eyes, Ron is a triathlete, accustomed to sixty-mile foot races and two-hundred-mile bike races. He met up with Tom Weis last year when Tom was preparing his bicycle tour of the entire XL route from Canada to the Gulf of Mexico. Ron rode most of the route with Tom and acted in a supporting role, reconnoitering food and lodging, setting up interviews, and shooting video. When they met David in Texas, Ron stayed on to help David build the tree village and set up the Tar Sands Blockade.

Raised in Madison, Wisconsin, and Charleston, South Carolina, Ron moved after college to Missoula, Montana, to enroll in an environmental law program. But he decided against it; the academic life was not enough. He wanted to be in "front-line activism." Ron became aware of tar sands mining well before most people in the United States because of the shipment of huge 250-ton mining trucks through Montana to Canada. "The trucks wouldn't fit under bridges or power lines, and took up two lanes of traffic, so they had to shut down roads in both directions to let them through," Ron told me. "There were no bonds for highway damage, and they didn't care about inconvenience to locals. They wanted to bring something like three hundred of these trucks through. The first one got stuck. So you start to ask, 'What the hell are they doing up there in Alberta?' "

Ron Seifert became aware of the pipeline while living in Montana.

Ron became aware of the extent of the tar sands issue the same time I did, and he responded in the same way. In June of 2011, we got the e-mails from 350.org about the D.C. Tar Sands Action, the massive civil disobedience action at the White House. "I got on board with that right away," Ron said. "I had a feeling I wanted to be more involved. I was unfulfilled with what I was doing and overwhelmed with this urgency that this has to be stopped." He was arrested in Washington a few days after my own arrest. He was now the coordinator of the Tar Sands Blockade here in Winnsboro and would be lining up the opening battles of the struggle against the Keystone XL pipeline.

When I asked what motivates him to devote so much of his life to environmental work, Ron said that he was brought up camping and hiking and came to love the outdoors, but feels that environmental work is not a choice: "Everyone in my situation ought to be doing what I'm doing. We have to." He feels it is a "lucky coincidence" that he also has such a passion for the natural world. "The fact that I enjoy it is fortunate for me. I have to do it, and I want to do it."

"Stopping tar sands extraction is a necessary condition for the future of the planet," he continued. "The science is there. If we don't stop it, it will be 'game over for the planet,' as the cliché goes. With the upgrading, piping, and refining of tar sands, its carbon footprint ends up three times the size of conventional crude. That's just unacceptable. They're on track to destroy the planet if all that reserve is exploited. We will reach the thresholds where we won't be able to rein it back in. It will be too late. We'll get into a feedback loop that will continue to spiral higher and higher where even if the whole global economy stopped, it wouldn't make a damn bit of difference.

"I'm a student of philosophy," he went on. "I took courses on formal logic. So truly, this is logically a necessary condition for our future. It's not a *sufficient* condition—it's not the only thing we have to do—but it's a necessary one. It has to be done in concert with a whole host of other things. If we don't at least stop tar sands exploitation ... that's it.

"So for someone like me," Ron said, "relatively young, doesn't have property, doesn't have a family, isn't married, doesn't have *debt,* that has the ability to go out there and be a defender of this planet: I don't see that's really a choice. How many other Americans are like me—that fit those criteria and have the passion and can actually throw themselves into the fray? It's probably less than 1 percent, maybe one-tenth of 1 percent. But that's still thirty thousand people! We could stop almost anything with that many people. If you believe we have five years, or at most a decade, to dramatically change the economy that exists—the whole unfettered growth model—you've got to break it down into things you have to do, and one of those things is to stop tar sands exploitation."

"Are you optimistic or hopeful?" I asked. The dark was descending as we spoke, but it was still hot. There was screening on two sides of the tent, but no air moving through. A screech owl cooed in the distant trees.

"I call myself a *pathological optimist,*" he said. "I will seek out the sunlight and the silver lining where it exists. But there's little reason to be optimistic. We have all these public interest lobbies and big NGOs [nongovernmental organizations], and they've done a great job mitigating the rate of loss. But the bottom line is we're moving toward the destruction of natural systems, we're not moving toward sustaining those systems."

"I think we don't have the worldview to do that."

"Exactly right."

"So, what have you and David been working on here?" I was hesitant to ask, because the existence of the tree village was not yet generally known. Ron and David wanted to delay TransCanada's knowledge of their plans for as long as possible in order to delay whatever counter strategy they might come up with. Construction of the pipeline will be stopped here: there is no question of that. The question is, how long will it take TransCanada to figure out a way around, under, or through the tree village?

"While they're busy rerouting easements through Nebraska, they don't want to be aggressively attacking landowners in Texas," Ron said. "That's in our favor. They will probably just try to wait us out. They could bring in some cherry pickers or cranes, but I don't think they can get that kind of equipment out here. We've constructed 'squirrel guards' to protect the tree houses from climbers coming up the tree trunks. I've nailed chicken wire to any branches they could throw a rope over. The only option they have to attach rigging is at the ends of branches, where it will be too dangerous to climb on a rope. It's a huge deterrent. They may try to bring in professional extractors or a rescue team of some kind to physically remove the sitters, but it will not be easy. We're ready to stay up there as long as it takes. We may surprise them."

"Are the sitters prepared to do that?"

"I think so. David certainly is. The other four haven't done a tree sit before."

"Will you be sitting?"

"No. I will be the ground support."

"Have you talked to Julia Butterfly Hill?"

"Yeah. She may come here."

"That would be fantastic. Have you thought about rotating the sitters?" I asked. The supply tree is outside the easement area and connected to the other tree platforms with a cable trolley, so sitters can come down from their trees using the trolley without setting foot on the easement. "People could even come down at night for a break or during a storm," I said.

"That's another option," he said. "If they try to wait us out, or try to leverage David legally, we think we can rotate more sitters in. We can keep supplies going to any number of people. So, it's just a matter of how extreme the reaction is. This isn't just mischief; this is a long-term strategy to stop the pipeline." A gust of air pushed through the trees above us and opened the tent flap. Ron stood and closed it.

"They could try other things, too," he said, sitting back down. "They could under-bore [dig below] the whole property, like they do under roads, but that would be crazy expensive for them. They could reroute the pipeline, but the whole process of acquiring land would take a long time, and there's a wealthy lakefront property to the west of here, so that would be a hell of a fight. I think the cheapest and fastest way would be to get us out of the trees. So, I won't be surprised if they hire a Blackwater team or something. Or maybe get the National Guard or state troopers out to section off the area and cut off the ground support."

"When do you expect this all to begin?"

"They say they can lay between two and three miles of pipe a day," he said. "If they can, and they actually start near Paris on August 1 and work south, they will get to Susan [Scott]'s place [where the Tar Sands Blockade training later took place] on about the thirty-fifth to fortieth day, and here on the fortieth to forty-fifth day. That's mid-September. So we anticipate Susan's place to be first, and we're going to portray the action there as the end of the Tar Sands Blockade. People will know something is going to happen in Winnsboro, but we would just as soon the attention went to Susan's place for now."

"So you don't want too many people to know about what's going to happen here."

"No, the element of surprise will be important. If they know what we're doing here, they will be able to respond sooner, and we will not be able to delay them as long."

July 23, 2012 – Winnsboro, TX
High 91°F – Low 75°F – Precipitation 0.0 inches

Ron and I left the campground at David's this morning and drove twelve miles into Winnsboro. Ron wanted to check out reports of surveyors in the neighborhood, while I spent some time getting to know Susan and Gabriel Scott at their home three miles east of town.

"WE'RE IN TUNE WITH NATURE, LIKE A BLADE OF GRASS HAS got feelings. That's kinda how I think," Susan began. "But I'm not any kind of religion; I'm a free spirit. I've always had the feeling that I was part of the earth. I've always been the protector, the giver, the Mother Earth, whatever you call it. God is infinite ... It would take me a lifetime to explain."

"Well, just take maybe an hour of your lifetime ..." I said.

"I think we're the keepers of the earth, and this is all we've got. This is it. We don't get to ride on a damn spaceship and go to another one and screw it up. We're supposed to be here to maintain our own self-contained spaceship, okay?"

"Good planets are hard to find," I added.

"And if we don't have enough sense to fight for and protect it, we don't need to be here."

"We need money, we need an economy, but that comes second," I agreed.

"Here's my opinion," she said. "Money is the means to get what you need. It's fine if you got it; it's fine if you don't. You can't eat it, you can't plant it and grow things from it, you can't breathe it, it doesn't bring you fresh air like the trees do. It's a thing you have in your hand to give somebody to get something in return. But this Earth is going to be destroyed because of it. I *feel* this strongly."

Susan and I were sitting in rockers on her front porch, enjoying the artificial breeze from an old, rusty electric fan. She bought it at a yard sale from an old man in his nineties whose parents bought it sometime when he was a child. Susan is a collector. Bird feeders, wind chimes, washboards, dipper gourds, and log chains hung from the rafters overhead. Road signs nailed to siding and fence posts said Bridge Out and Authorized Personnel Only. Susan is a great-grandmother, strong and vibrant, with a friendly smile that says, "You're just plain folk like me." Her son, Gabriel, stepped out from the kitchen to join us. He was in a Signal Corp Battalion in Desert Storm working at Patriot missile installations: "Scud busters, we called them." Gabriel has two granddaughters.

As we were talking, Ron Seifert suddenly pulled up in the driveway and rolled down the window. "The truck's gone. I imagine they headed up to Clearwater Highway," he said. Susan had seen surveying equipment that morning set up next to her fencerow along Route 11. She called David and Ron right away. "I might have to run up there and see if I can find that truck," Ron continued. "That will give us an idea of where they are."

"Yeah," Susan agreed. "If they're up there, you know they're fixin' to come down across Youngman."

David Daniel, Ron, Susan, and Gabriel knew they could not stop the surveyors, but they could delay them. They planned to make a stand at the border of her land and challenge the surveyors, firmly but politely, as they tried to cross. Members of the survey crew were just doing their jobs—they were not the target—but they would have to back off and call the home office, or go into town and get the sheriff, or a court order, before they could enter. This was not to be a civil action, just a delaying tactic. Nobody was going to risk arrest at this point. "I'll call if I see anything," Ron said as he backed his truck around and disappeared in a cloud of dust.

"Okay, we'll meet you wherever you say," Gabriel shouted. "David is heading this way, too."

Gabriel sat down with us as Ron drove off.

"I've got this little voice recorder on, Gabe. So be careful what you say," I said jokingly. "I'm not going to put you on the radio or anything."

"I don't care if you put us on TV," Susan said. "I don't care if you put us right in front of the president of the *U*-nited States. He needs to know. I think he's against it. I think he got bulldogged. He got intimidated or bought out, one of the two. That's my opinion. He's wishy-washy.

Susan Scott and her son, Gabriel, are taking a stand against the pipeline.

"I got goaded into signing by fear alone," Susan went on. "Pete Porter, the land agent for TransCanada, was a slick-talking man. He could talk the horns off a billy goat and sell them back to him. He said they're coming, they're going to take my land, and if I do anything they would sue me. I was going to have to pay lawyer's fees, court costs, and everything else. This farm is a natural land. We have big trees. This is what we live by. We're grateful to be here."

"It's your home. I can relate to that," I said. "I have a homestead in Kentucky. I chose to be there because I wanted to be surrounded by the natural world, and I feel for what you are going through here."

Gabriel sat forward. "I think we're just going to show everybody how peacefully we can tell them we don't want them here, without violence. We're not violent people," he said.

"Your role here is going to be absolutely essential," I assured him. "You won't be the only people the media talk to, but they will want to hear what you, the landowners, think about what is going on."

"I ain't real good at talking," Susan said in a low, dejected tone.

"You should just speak from the heart," I said. "People will want to know what you are feeling. As you say it, it will just come out. But like Gabe was saying, how people understand this will be so important. It's not about the confrontation. It's not just us against the cops."

"It's about the earth."

"Yes."

"You take water," she said. "They use eight barrels of pure drinking water to clean just one barrel of that stuff [tar sand]. Then they dump it in a tailings pond that birds land in and die."

"We're not radicals," I said. "We're just trying to let the earth speak through us. I'm sure you can do that with composure and calm. I know you will be firm."

"You gotta stand your ground," added Gabriel. "You know I'm a Desert Storm vet. They sent me to fight for oil; I truly believe that's what that was about. But when I came back here, I just wanted to relax, to enjoy my life here. I have a two-year-old granddaughter we're raising, and I was looking at their map the other day, and I see this big red line going down through the middle of our land, and I said, 'What is that?' It's what they're calling a *kill zone.* Even if there was a pinhole in that pipeline, at 1400 psi, that's like a bullet."

"And they won't be able to detect it," I said.

"Like the Kalamazoo leak. They thought it was an air bubble in the line; they *increased* the pressure to push it through," he said.

"What's your understanding of what's going to happen with the training this weekend?" I asked.

"My taxes are paying our government to protect us," Susan answered. "They've bailed on us. These politicians come shake your hand, but when you want to talk to them, you can't touch 'em with a ten-foot pole."

"These kids coming—I call them kids," Gabriel said. "I'm forty-six years old. A lot of people would look at them and classify them as misfits, just by their appearance. But there is a lot of knowledge there, a lot of wanting for the earth to be around for their kids. I see that in them. They're smart. Everyone I've talked to is smart."

"Do you trust them to express your interest?" I asked.

"I do. I do trust them," Gabriel assured me. "They're face-to-face with you. They'll look you in the eye. My grandpa said if you shake a man's hand and he don't look you in the eye, you need to look somewhere else."

"They're trying to save the earth," Susan added. "They're going to be camping out here."

"And if they're doing something you don't agree with ..."

"They'll have to leave," Susan said firmly.

"They have to speak for *you*," I agreed. "I think this is crucial, what's going to happen right here. Stopping this pipeline is critical for the earth's entire system, and you guys are going to be the first ones. You will set the tone for what happens elsewhere."

Susan was looking out over the porch to the woods. "You know, I try to talk to people I know about this, I try to get through, and even the trees answer better than they do," she said. The air was hot and still beyond the fan's range. The birds were silent, and tiny black butterflies fluttered through the weeds along the dirt road in front of us.

"I understand that," I said, after a long pause.

"And all the dirty air they make burning that stuff. You can't put it in a sack and tie it up. And the water, too."

"If there's no water, then there's no us," Gabriel added.

The phone rang, and Gabe got up to answer it. It was Ron. No sight of the surveyors.

"You know, your heart breaks in two," Susan was saying in her soft, dejected tone. "Literally—I had a heart attack over this. I never took no medicine. Now I have to take pills every day. I don't like conflict. I'm the most peaceablest person I know on this Earth. I'm plain country."

THE THREE OF US WALKED DOWN TO THE POND WHERE THE encampment would be. Two large tents were already pitched and ready for whoever would come. A swarm of large channel catfish boiled the water in front of us as Susan filled a bucket with fish food.

"They can hear us coming," she laughed as she threw a handful of pellets out to the hungry horde and the feeding frenzy began.

"Yup," Gabriel said. "My granddaughter just loves this lake. We fed the cats yesterday, and a big one turned and splashed us all over. She will never forget that!"

July 24, 2012 – Winnsboro, TX
High 93°F – Low 75°F – Precipitation 0.0 inches

Susan found the surveyors near her land this morning and called Ron. He jumped in his car and made it to Susan's place in time to video the encounter and post it on the Stop Tar Sands website. Susan had a tremble in her voice at first, but she seemed to gain confidence as she spoke to the surveyor. He was polite (the camera might have helped) and agreed to stay off Susan's land until the situation was clarified. David and I showed up as the discussion ended. She smiled her "just plain folks" smile to the crew chief and seemed to relax. I thought she did very well.

We also heard today that the Army Corps approved the middle section of the southern leg, the section that runs through Winnsboro. It's full speed ahead now, from Cushing to Port Arthur. If they keep building north to south, they should get to Susan's land in early September, and the twelve miles south to David's a week or so later.

IT WAS A FEW DAYS BEFORE I GOT A CHANCE TO SPEAK WITH David Daniel at any length. He had been busy preparing tree houses, caring for Clara and Naida, and coordinating events at the training that weekend. The Keystone XL pipeline had been a strain on his emotions, his family, and his finances. He is a quiet, shy, soft-spoken man, but utterly determined to make a stand, right here, on his own land. Uncertainty surrounds every aspect of his life, yet he knows exactly what he is doing. I knew, well before I came here, that the early tone of the movement against the pipeline would depend largely on his character, and I was uncertain as to the maturity and seriousness of that character. I did not know how long I would stay here or how involved I might become in the blockade. But after meeting several times with David, I decided to stay through the training.

We walked to David's home just up the hill from the campground where Ron and I were staying. David's a carpenter, and like me, built his own house. A few unfinished touches remained, as in most owner-built homes, but the space inside was airy, light, and tidy. Slate and ceramic tiles lined the floors, and salvaged tin roasting pans covered the ceiling in the kitchen. They gave a well-sculpted, homemade look to the room. A foot or more of insulation filled the space between ceiling and roof.

David began talking about his legal struggles. "Eminent domain is a tool industry uses, a big tool, a tool that is the destruction of our rights and our use of the commons," he said. "Right now, industry has a free-for-all. All companies have to do is go to the Texas Railroad Commission and get the one-page T4 form permit and literally check a box that says 'Are you private, or are you public?' If they check *private*, they have no right to eminent domain; if they check *public*, then they have eminent domain."

"They just have to *say* they're public?" I asked.

"That's all they have to do. The railroad commission has no oversight to check and see if the company is what it says it is. It's a self-declaratory process by the industry."

"So this has never been tried in court. The definition of public use ..."

"There has been a recent case, the Denbury Green case," David said. "The Texas Supreme Court said that the T4 form permit *does not* prove that a company is a common carrier for public use. So a landowner has the right to make that challenge, but do you have the *means* to make that challenge? Do you have the million dollars it would take to bring it before the Supreme Court? Ultimately, the way I see it, eminent domain is a big tool that industries can use to begin their destructive projects that threaten our resources ... and our rights."

"You got into this as a landowner," I noted.

"Four years ago this month, July 2008, my neighbor called and said people had been trespassing on the property. I got home and walked the property, and I saw that it had been fully surveyed and staked, cutting my land in half. There were all these orange stakes through the big hardwoods down by the creek that said 'XLPL 36 inch.' I didn't know anything about TransCanada, the Keystone XL, or anything about the proposed project."

"So they came without your permission, or even notice, and staked out your land, trampling right on top of your home?"

"Yes. The 'PL' on the stakes led me to believe this was for a pipeline. A month and a half later, I got a letter from TransCanada. Only then did I know who the stakes belonged to."

"In that month and a half ..."

"I had no idea."

"What were you thinking?"

"My heart sunk," David said. "The area that was staked was the reason we bought the twenty acres, because it has the springs and the creeks, and the old-growth hardwoods. That's where these stakes were. The letter from TransCanada had a phone number, so I called to ask why the surveyors had come on my land without permission, and the guy said 'They probably got lost!' "

"Ha!" I laughed. "Surveyors don't get lost. That's their job!"

"Yeah, with GPS and all—I didn't buy it. I got another letter in November," he went on. "They said they had the right to eminent domain, and if I didn't comply with their requests within ten days, they were taking me to court."

"Did you talk to your neighbors about it?"

"Nobody knew anything about it. We all thought it was a gas pipeline."

"Did you know then that they would be cutting down your trees?"

"I assumed they would have to cut some trees, but the scale and scope, I had not a clue."

David thought for a while, as if feeling again what he had gone through. "After I got the letter threatening eminent domain, I stayed up into the wee hours of the night composing an e-mail saying who I was and what my concerns were, with a series of questions about what they intended to do. I got a phone call the next morning from an attorney saying he would forward my questions on to Houston. [David heard back on his list of questions nearly two years later.] Then the attorney said, 'All I need to know from you is which pile to put you in: the compliance pile, or the f–cking uncompliance pile."

David Daniel is fighting TransCanada's eminent domain seizure of his land.

"He said that?"

"That's what he said." (David told the same story on a radio interview with NPR. According to the interviewer, the TransCanada lawyer later denied using those words.)

"Really. So he was already contentious."

"Yeah, in a *you're-a-peon* sort of way—I am a peon up against a multinational corporation with lawyers that would just eat me alive. I can't afford an attorney."

"When the surveyors came again, what was your interaction with them?"

"They were just people doing their jobs. They're not responsible for all this."

"That's probably true of the construction people, too."

"Sure. It's just trying to keep that in perspective. Those are the faces you see that are doing the trespassing," he said.

David began researching the tar sands in early 2009, though there was not much on the Internet that long ago. But he soon learned the difference between tar sands and conventional crude oil. "There's heat involved in a tar sands pipeline, and much higher pressure. The pipe is only 0.465 inches thick. I've done a lot of welding and worked with steel—that's pretty stinkin' thin for a high-pressure pipeline. Combined with the heat, combined with abrasion from the quartz sand particles. The high pressure has to do with the nature of the bitumen. Tar sands bitumen is virtually a solid at room temperature. Conventional crude oil is a liquid; it flows easier. Bitumen has to be pressurized and liquefied with carcinogenic diluents to flow. They pipe it at 140 to 160 degrees. Originally, I thought they were heating it to make it flow, but it turns out the heat actually comes from the friction."

"So that's got to be abrading the inside of the pipe," I added.

"Yeah. To me that seems common sense," he said. "Later on, when we had our [Winnsboro Stop Tar Sands group] meetings, we'd have oil field workers and pipeliners showing up asking why we were opposing the new pipeline. I gave them the information about the pressures, specific gravities, and heat, and they started saying, 'Wait a minute, this doesn't work!' So we had people who actually built these things express concern."

"So, this was not going to be an ordinary pipeline," I said.

"Exactly."

Like many others, David's environmental consciousness developed as he came to know more about the tar sands. "After I started reading the environmental impact statements, I began to realize how toxic the bitumen by itself is. And then they add all these other chemicals. As much as 1.7 million gallons a day can leak from the pipeline without triggering any of their leak detection systems, and the leaks will be underground. They could be leaking into our water supplies without anybody knowing about it. The Keystone 1 has had fourteen leaks since it became activated, all *above* ground, where they could be seen. What's happening below ground, we don't know. We're sitting on top of the Carrizo-Wilcox Aquifer, which feeds water to ten to twelve million people in Texas, and on top of that is the Queen City-Sparta Aquifer, which is only a few feet underground. I drink out of the springs on my land now, but I won't be doing that anymore."

"You don't know what you're drinking."

"And testing your water is expensive," he said. "That's the landowner's responsibility. After all the negotiations, I got them to agree to test my water one time and had to fight for that much. And they threatened to take that away from me if I didn't sign with them. 'If you don't take this offer,' they said, 'we'll take you to court. You'll have to hire an attorney and appraisers. So what's it gonna be?' "

David has always been concerned about the environment. He was a Boy Scout and has done a lot of camping, but he never belonged to an organization until the pipeline came along. "I guess I had my head in the sand and my fingers in my ears, but that's no longer the case. What it took for me was to have it come banging on my door. Now I have a front-row seat."

"I'm a father," he went on, "and I have a responsibility to my daughter, and that's not just financial, that includes what kind of environment she's going to have to live in. The World Bank has stated that the next wars are going to be fought over water, and the climate implications are right here in front of us. Certainly, the earth goes through cycles, but we're speeding the climate process along. We have the knowledge as a species to have a positive impact on the environment, but we choose to have a negative impact. This planet we're floating through space on is all we have. It's like what I heard a tribal leader say, 'The earth is sick and has a fever. It's a living thing; when something is sick and has a fever, you take care of it.'

"But you can't use the word *environment* around here without, in some cases, risking your safety," he continued. "What we've done here with Stop Tar Sands is learn to speak a language that people around here understand. I can still walk through town, and I don't have to worry about anything."

"That's going to become more important," I said.

"I believe in the environmental message, but the media still turns that message into the environment versus jobs and leaves it at that," David said. "So I want to come at it through all the other doors that have worked, and get the support of people who normally wouldn't support an environmental message. I'm a little concerned about other people and other groups getting involved who don't understand the local culture—people who don't speak the local language. I've heard rumors that people might be skydiving in or do some other sort of showboating that would not go over well here."

"So the environmental message should be around the periphery of what you're trying to do. Is that right?" I asked.

"The environmental message is the elephant in the room. Everyone will see it. Every way into the room leads to that message. And people should know that we did not start out with any sort of protest or civil disobedience. We've gone through and exhausted every possible avenue that you're supposed to go through, and this is what we're left with."

David, it turns out, is a trained gymnast and "ran away with the circus" at age 28. That will prove important for what follows here in Winnsboro. He spent twelve years rigging and performing high-wire circus acts all over the United States. "The real work is in setting up the rigging; performing is the easy part. That's when you can relax," he said. David knows what he is doing with cables, pulleys, and tightrope acrobatics. "When I rigged a show, I had to be thinking not only about my own safety, but the safety of the spectators and other actors."

TransCanada might just be messing with the wrong Texan.

THE NEXT MORNING, DAVID SHOWED UP AT THE CAMPGROUND and asked if I would like to see the tree village up close. As a solar installer, I've had some experience with heights and had no hesitation. He led me first to tree number one—the supply tree, strategically located just *outside* the easement. There was a "bunk" platform built about ten feet from the ground, up a wooden ladder.

David handed me a harness. I climbed in and adjusted the straps to my waist and legs. "You have a carabiner [shackle] at waist level, and another one at the end of this lanyard," he explained. "The waist carabiner is for the trolley; the one at the end of the lanyard is your safety attachment." We climbed up the wooden ladder.

"This platform is for the media," David told me. "The 'primary platform' is up there." He pointed up a thirty-foot extension ladder leading from the bunk up to the first big branches of a large red oak tree. "There will be ground support from the base of each sitter tree within the easement, but the supply tree can get sitters and supplies to all five stations from outside the easement, if need be."

"It's still pretty brushy at this height," I noticed. "Will the media be able to go up any higher?"

"If they want to," he said. We climbed the extension ladder, one at a time, to the primary platform. From there we hooked in our lanyards and climbed another twenty feet up a cable ladder to the "launching platform," where the trolley was located. A cable loop dangled down from two pulleys on quarter-inch steel cables. A three-quarter-inch manila pull rope was independently suspended along the side. An additional smaller rope was attached to the trolley to pull it back empty to the first tree, like a line of laundry hanging out to dry.

"If one cable breaks, the other will hold, and there is still the manila pull rope," David explained. "They're rated for several thousand pounds. After you attach your waist carabiner to the trolley, attach your lanyard to the pull rope for additional safety. It will slide along with you."

"So if the cables break, I'll be dangling from the pull rope halfway between two trees with my feet kicking the air," I observed.

"Yeah, we'll get you down," he laughed. The next tree, number two, was about eighty feet away. "When you're hooked in and ready, just pull your feet up off the platform, put the pull rope under your right arm, and pull! I'll go first." David snapped into the trolley and easily pulled himself over to the next tree. The trolley came back with the help of the smaller rope.

I snapped in. "I love you, dear!" I shouted, as I leaped off the platform. "I love you, Bonnie! You can have all my stuff!"

David's tree has a the trolley landing at the bottom and a sleeping level on top.

I never felt uneasy as I coasted through the canopy to the other side. The pulleys turned easily, and I felt no physical strain pulling the rope. I stepped off on the launching platform of tree number two. This was to be David's home for the foreseeable future. Chicken wire was nailed into the base of all the larger branches so that no one could throw a climbing rope over. Ten feet below was a "squirrel guard": sheet-metal roofing on a wooden frame extending out from the tree to prevent climbers from coming directly up the trunk. Anyone with a bird feeder knows how a squirrel guard works. Another trolley would launch from where we were to tree number three, with two sitter stations. But we were going higher in David's tree.

"We're way too high up here for them to use a bucket truck or any kind of crane they could get in here," David said. From the launching platform, we climbed further up a rope ladder to a covered four-foot-by-eight-foot shelter. Inside were water jugs, food, sleeping bags, a compost toilet, rain gear, tents, and solar chargers—enough for a month in the tree without resupply. I wondered where he would sleep. We could see tree number three easily, and number four beyond that. Number five led from three, but was barely visible from where we stood.

"There are no trolleys past tree number three, only ropes. But it's still safe," David assured me. "Sitters in tree three will have to go through trees one and two, and sitters in four and five will have to go through three. But we can visit back and forth, and rotate sitters in and out. Guy wires connect all these trees, reaching out to fill a 120-foot-by-120-foot square, spanning the width of the easement, so it will be impossible to cut any one tree without tangling all of them. If they keep this route for the pipeline, they will have to pass through these trees, and I don't know how they are going to do that. They can't cut any of the guy wires without endangering us."

"That would be bad PR," I said.

"And bad karma," he added.

I had seen enough to get the general idea and did not insist on moving over to the next tree, but David motioned me out the entrance of the shelter. "Hook in here," he said, and he started further up the tree to the roof of the shelter. I followed him up through the branches and hooked back in. Now we were close to the top of the canopy, about eighty feet in the air. The roof was a little larger than the platform below, four feet by twelve feet, with a pole for a tarpaulin running horizontally along its length.

"This will be a tent for my sleeping space," David said. "Over there is the escape pod." I was still taking in the view across the treetops as he pointed to a tiny three-foot-by-four-foot platform suspended by its corners from another tree about sixteen feet away. A cable connected the two trees. "If any extractors make it up this far, I can cross with three or four days' worth of supplies and tie myself in. But the platform is very unstable; it will swing wildly if anyone else tries to come over. This is the last ditch, but I will have a rope if I need to get down for any reason."

"The tree house we're standing on, by the way, can sway independently of the tree. It's stabilized by the guy wires."

"Is this whole system safe?" I asked, a little too late for effect.

"I question anybody who doesn't have that question. Everything I do in rigging, I do overkill. But you always have to remember where you are."

The tree village took three months to build. David built the platforms on the ground. Ron came along at just the right time to help him carry materials through the woods to the trees, hoist them up, and assemble them in the branches.

THE NEXT DAY, I WAS SITTING ALONE IN DAVID AND CLARA Daniel's woods, down near the creek. The cicadas were singing their summer song. Squirrels were scrambling through the branches and blue jays scolding in the distance. There were tall hardwoods here, like the ones I know back home: oak, hickory, gum, and maple. A red oak tree, thirty inches thick, towered a hundred feet overhead. This is the western edge of the great North American deciduous forest that I have come to love, to feel at home in, and to feel called upon to preserve from the biocide of mountaintop removal. This is where I live. It was hot but shady; a soft summer breeze stirred the limbs overhead. High above were platforms stocked full of food and water, now empty of people. Seven feet below ground was the future bottom of the Keystone XL trench, now bedrock and undisturbed soil. This was where the earth will rise and make her stand.

As I sat, I wondered what I would say if, like Susan, I could speak to the trees. How would I explain to them why I was here, and what we humans were up to? Would they see me in my two-legged uniform—an arm of humanity—or would they distinguish me, with Ron and David, Susan and Gabriel, from others of my kind? Would they hold me responsible for what was to happen? Would they see with eyeless branches, or only feel the communal presence of my breath? Their days were numbered, yet they let me sit there, undisturbed, under their branches and over their roots. They did not seem to mind. If I could speak to them, would they reply? What would they say to me, and to my kind?

A woodpecker called.

Did they know they would die? "Trees don't die; the forest dies," I heard, but it was in the thoughts and words that hover over breath. In seeing, there is separation in space; in breath, there is no space. I was trying to listen, but humanity held me. Words clouded my mind. I was listening through the tangle of English; they spoke no English. They breathed: in and out, and up and down. I felt them pushing gently on my lungs—the other side of sensation. The earth pulled down and held my body up. Is that for granted? Will it always hold me here? The earth spoke. Yet, again, I heard no words. There were no words. I listened to breath, to gravity, but did not know what they said. There was no doubt as to their being here, with me. There was no question as to reality. I wished to learn their language, to calm my thoughts, but then, could I speak to people? Would they understand me? Are there wordless words for people?

The woodpecker called from a distance, a quiet trill.

I sensed the forest retreating to low places like this. The tall oaks are dying on the hillsides, parched from last year's drought and this year's. The heat and dry Earth are too much for them. The deer, the panther, and the armadillo will have less space to roam. Do the people who build pipelines breathe the quiet call of the earth? Do they sense the forest? Do they know they are piercing a stronghold whence the trees may one day spread anew? Does the life that is here show on their maps of towns, roads, and property lines?

The woodpecker called again—this time in a loud, raucous squawk. His home is the old trees that have had their life. He lives where I live. I stood, a piece of my heart in David's woods, waiting for the chainsaws and bulldozers.

July 27, 2012 – Winnsboro, TX
High 95°F – Low 78°F – Precipitation 0.0 inches

The Carbon Tracker Initiative, a group of environmentally oriented financial analysts, published a new number representing the total amount of carbon in the proven reserves of oil companies and oil states such as Kuwait and Venezuela: 2,795 gigatons. This is the carbon that will become available on the world market in coming years. One-fifth that amount, 565 gigatons, is the maximum amount of carbon we can burn to stay within two degrees of global warming, according to the current scientific consensus. That means we will have five times more carbon available to us than we need to ruin the climate!

Bill McKibben writes in *Rolling Stone*: "Think of 2 degrees Celsius as the legal drinking limit—equivalent to the 0.08 blood-alcohol level below which you might get away with driving home. The 565 gigatons is how many drinks you could have and still stay below that limit—the six beers, say, you might consume in an evening. And the 2,795 gigatons? That's the three 12-packs the fossil-fuel industry has on the table, already opened and ready to pour."[25]

This confirms what I read earlier in Leonardo Maugeri's *The Age of Oil*. We will not kick the habit by running out of substances to abuse. We will have to do it consciously. We will have to dip into the practical realm of consciousness, the tool unique to our kind. Survival will be a decision.

OUT OF NOWHERE, I GOT A CALL FROM TRANSCANADA. WELL, almost nowhere. Planning for my Alberta trip a few weeks ago, I placed a call to what I thought was TransCanada's Fort McMurray office and left a message asking for an interview. As I hung up, I realized that the call went not to Fort McMurray but to their main office in Calgary, where I would not be going. It wasn't going to work, so I forgot about it. But they didn't. So here I was, talking on the phone to Don Wishart, TransCanada's executive vice president of operations and major projects—basically, the guy who is building the XL through David's woods. I began with the easy stuff.

"We hear a lot about jobs. How many people will be employed building the XL?" I asked.

"At the peak of construction we will have about four thousand people on the job site, around the fourth quarter of this year," he told me.

"Is it too soon to say how many permanent positions you will have available for operating the pipeline when construction is finished?"

"It is certainly not a labor-intensive industry," he admitted. "It's a capital-intensive industry. We generally would have a team of technicians and technologists located at about every second pump station. So, in that area [the southern leg], we would have three or four teams of electricians, instrumentation technologists, and mechanics located along that particular piece of the route. Engineering and more sophisticated support would be in Houston, where we have a large office."

"I've talked to a lot of people along the route who make a big distinction between tar sand bitumen and conventional crude oil. Is there any difference in how the pipeline will be constructed and operated considering that it is a nonconventional form of crude oil?"

"The short answer is *no*," Don said. "It is becoming *the* conventional oil. Heavy oils are on the upswing in terms of the supply of oil around the world. The U.S. has produced very heavy oils out of Bakersfield dating back to the early 1900s. Because the viscosities of these oils are high, they tend to be blended with low viscosity hydrocarbons, condensates, or lighter crudes so that they will move in a pipeline. Once it [bitumen crude oil] reaches a lower viscosity, it behaves like any other oil. That's all it really is. It's just a heavy oil; heavier than your lighter, sweeter oils."

"Bitumen oil, as I understand it, flows at a high temperature. Is this because it is heated, or because of friction in the pipeline?"

"It is not heated," he said. "It is just friction and pressure. It's not much different than any other oil, or even natural gas in a pipeline. We have to deliver oil to the refinery at a temperature not to exceed a given level, I think about 60 degrees Celsius."

"First responders I have talked to along the route are concerned about the diluents that are used to make this crude oil flow in the pipeline," I said. "They're concerned because they don't know what those chemicals are. They talk of benzene and naphthalene, but they don't know what else might be there. They are not even allowed to know what is in there; they're told it's 'proprietary' information. They feel that they are being kept in the dark."

"Crude oils are not refined products," Don said. "There are a lot of components in crude oil that when released are not good for the environment. You need to clean them up, and you need to do so aggressively and expeditiously. That's a fact. As the pipeline company, we do not control the composition of the crudes going into the pipeline. It must meet the specifications of the refiners who are our customers and the generally accepted criteria set by a standards organization. This allows the oil to be exchangeable, so we can move it from one pipeline to another. How you meet our criteria for viscosity, for instance, is somewhat your decision. Each batch we pump as a complete stream may be different from the one upstream or the one downstream. These diluents are generally hydrocarbon-based, and they are not good things that you would want to drink. But they are ultimately the things that we use to make the refined products that we use at the other end."

"In other words, as the pipeline owner, it's not your product moving through the line, you're just providing the conduit."

"That's correct, Sam. That conduit is available for people who meet certain criteria, but it's not every criterion, it's just like the viscosity and some of the chemistry aspects."

"But can you see that from the standpoint of the landowner, they don't have a relationship with the refiners or the extractors of the product; their relationship is with you. So, first responders are in the dark as to what they are responding to. With the Enbridge problem in Michigan, for instance, first responders were ready to skim the crude oil off the surface of the river, but the tar sands bitumen did not come to the surface; it sank to the bottom. So they were without any kind of preparation as to characteristics of the product running through the line and therefore did not know how to respond to it."

"I appreciate that, and as a corporation, we probably need to do a better job communicating around those issues."

"I would say so."

"There's nothing absolutely, totally toxic, like cyanide gas," he assured me. "But if they caught on fire—which doesn't happen very often—that would be heavy black carbon smoke that would not be good for breathing in. It's not containing something that with one breath would cause irreparable damage. Clearly, they are not good for soils, and clearly, they are not good for aquatic environments."

"That's a major source of concern for people along the line," I noted.

"It's a very legitimate concern, and one that doesn't ever go away," Don said. "We appreciate that. We have been in business for sixty years. We placed a pipeline in an easement sixty years ago, and we still have a relationship with those people."

"Many people now are concerned that their easement agreements require them to be responsible for the pipeline when it comes to the end of its useful life." I was thinking of what Paul Mathews and Joe Moller had said about their agreements with TransCanada.

"That is a misconception. We are responsible ... from cradle to grave. Anyone who has suggested otherwise is spreading false information."

"I don't think anyone is spreading false information," I said. "That's the way they understand it, and it hasn't been explained to them otherwise." Then I changed the subject. "Have you found a new route around the Ogallala Aquifer?"

Don said, "We're working with the Nebraska Department of Environmental Quality to identify another route. Our routes are clearly designed to avoid the Nebraska Sand Hills. We were given a very specific map by the state and local environmental officials ..."

"That was the EPA map ..."

"But we can't avoid the Ogallala Aquifer. It covers six states. It's a very large feature. We can avoid the Sand Hills, we can avoid sensitive areas, but we can't totally avoid the Ogallala."

"Do you have a timetable on the northern leg?"

"We hope to complete the Nebraska state process by the end of the year or very early next year," he said. "Then we anticipate receiving the presidential permit within the first quarter of [2013]. In reality, we will start construction in the spring. You can't really work on the agricultural soils unless you can remove the topsoil. There are some things you can do—bores and things like that, drilling under roads and even some rivers."

"I have one final question: The Canadian tar sands contain enough carbon to permanently alter the climatic system of the planet. What role do you see this pipeline playing in changing the earth's climate?"

"That's a very good question," he said. "I will try to answer that in two pieces. There was a study in the journal *Nature Climate Science*. If all the carbon in the oil sands were combusted in one second, it would add 0.36 degrees Celsius to the world's temperature. They compared that to unconventional natural gas reserves—the shale gases and the hydrates in the oceans—and that would produce about 2.86 degrees Celsius change in the climate. And if you did the coal reserves on our planet and combusted them, they would change the temperature 14.8 degrees Celsius. So while it is a large carbon source, it is a fractional source when compared to others. The Canadian oil sands represent one-tenth of 1 percent of global greenhouse gas emissions. If they did double—that's the aspirational dreams of the producers—it would be two-tenths of 1 percent."

"That does not include upgrading," I added quickly, "which doubles or triples the carbon footprint of the oil sands." But I realized as I said this that, according to his figures, even doubling or tripling 0.36 degrees would raise the Earth's average temperature only about one degree Celsius—still enough to throw the climate off, but not nearly as much as I had been hearing from my own sources. Were Don and I arranging our own constellations in the night sky, or looking at completely different skies?[26]

"The determination," he went on, "was that there was about 6 percent more greenhouse gases from oil sands crude than from the median of other types of crude oil—only 6 percent higher than the average oil consumed in the United States today. Canadian oil sands produce about forty-five megatons of carbon per annum, which is about 3.5 percent of the emissions from U.S. coal-fired power plants.

"The second part of your question," he continued without pause, "would have to do with the role the pipeline would play with the climate. You need to start with the context of the United States as a consuming nation. The U.S. Energy Information Agency's 2012 annual report states that the U.S. consumes about fifteen million barrels of oil a day in the year 2012. With new renewable sources—wind, solar, all that—and huge improvements in efficiency, they predict in the year 2035 that the U.S. will be using about fifteen million barrels of oil a day."

"The same amount."

"It's the same amount. That's almost twenty-five years of GDP growth and population growth with no increase in oil consumption. So, it is actually a terrific outcome. It's a substantive improvement in energy efficiency of a nation. So it becomes an issue of, where do you want to be getting that fifteen million barrels from every day? If you choose not to source that from Canada, the likely source would be the Middle East. But if the pipeline is stopped, will that stop Canada from producing the oil? No. There is seventeen *trillion* dollars' worth of oil in the oil sands. They will ship it abroad. It will go to the West Coast; it will go to the East Coast. It will one way or another make its way to other energy markets, largely the Asian market. The implications of not having Keystone XL going forward and having that oil exported would increase the number of ships on the ocean by two hundred tankers per year. Other forms of transportation—be it trains, ships, trucks—are much more energy intensive than pipelines and would generate nineteen million more tons of carbon per year."

"So, it would be your assessment, then, that the climate is better off with the pipeline than without it," I said.

"Yes. I personally think that's indisputable. And despite the very legitimate concerns people have with pipeline safety, the alternatives are generally a full factor of ten, and sometimes thirty times more likely to have an incident than a pipeline would."

I started to say something about the $17 trillion—how it seemed we should get more than 0.36 degrees' worth of carbon for that much money—but thought better of it. This was Don's chance to speak. I continued with the flow.

"But with a pipeline, leaks are often undetected because they are underground," I said. "The pressure drop is not enough to be detected electronically. A leak can go on for a long time before it is known to exist."

"There are finite levels of detection capabilities," he agreed. "But generally the fluids will rise to the surface. With not very much oil, you will end up with a stain on the surface of the ground. We fly the line on a very frequent basis, and we are always looking on the ground for those areas where you can see a yellowing of vegetation. They are detected when they occur—they're not very frequent—and they're small volume: a barrel or two, and we are able to go in and remove those soils and replace them with clean soils."

I asked Don if he had anything else he would like to add.

"A lot of people perceive those of us in the oil patch as an evil and greedy group of characters," he said. "My background is in environment and in farming. I ran an environmental consulting firm for a number of years. I care about farming; I care about the environment. The reason I went into this work is that I was a young, wild-eyed person of the sixties and seventies that wanted to save the world. I work here for a pipeline company because these guys actually do care about this stuff. We're serious about being as responsible as we possibly can. We know this is our social license to operate. The resistance we've had around Keystone has been remarkable. I'm very sympathetic to most of the people who have these concerns. But we're not just a pile of guys trying to drive a pipeline through somebody's property and walking away. There are real people living in these communities. We still intend to be a neighbor sixty years from now."

"Where was it you were farming?" I asked.

"Southern Alberta. My father-in-law started as a homesteader and ended up with ten sections."

"Ten sections? That's a lot of land. Now that you are using kilometers, when you say 'sections,' do you mean square miles?"

"Yes. Square miles. We got halfway through metrifying this country and then stalled."

"But you still use metric dollars up there. That's the hardest part for us."

He laughed out loud. "But I like the different colors of the bills."

"Thank you for your time and your contribution to this project," I said.

"You are welcome," he said earnestly. "We are at your disposal if you have additional questions."

LATER, I VISITED BECKY AND SAM, NEXT-DOOR NEIGHBORS TO David and Clara Daniel. The pipeline will cross their land, too. They do not view TransCanada as positively as Don Wishart.

"We're not tree huggers by any means," Becky said. They laughed together, as if sharing another of many light moments at the expense of long-haired environmentalists. The pipeline will come down from the north and cut across their land as it will David's.

"But we're being run over," she quickly added. "Our fears, our concerns are just laughed off. You say to them, 'Please, give emergency contact numbers, please let me know what first responders need to know,' and we're told, 'We can't tell you that because it's against homeland security,' and I quote that. How are we supposed to live with that thing next to our house? A pipeline's one thing. If you live in Texas, you've got a pipeline across your backyard if you've got more than an acre and a half."

"Or an oil well," Sam added. "That's understood. But this tar sands crap they're trying to shove down our throats—why don't they run this stuff across their own country? I don't like the president we have now, but I respect him because he put a stop to most of this [the northern leg]."

Becky is more firmly anti-Obama. "We don't expect Canada to care about us. What we do expect is everyone from the local little guy right straight up to the man who sits in the president's chair to protect us. We're the ones they're supposed to protect first, not the fifteen million dollars John Boehner stands to gain from this. He's not representing us."

"Are you opposed to Obama's approach to this?" I asked Sam.

"I'm opposed to *Obama*, thank you!" Becky interrupted.

"He's a *post turtle*," Sam added.

"Post turtle?"

"Yeah. You're walking along the side of your property, and on the fencepost you see a turtle just sitting there. His legs are flopping every which way, and he doesn't know what to do. He couldn't have gotten up there on his own, and he can't do nothing once he gets there. He's a post turtle." But then he added, "We'll vote against Romney, too. He's all about big money."

"I think the government is too big for itself," Becky began. "They all need to go, and government needs to get pared back down to what the country needs so that something can get done that benefits everyone, from the top to the bottom."

The surveyors trespassed on Sam and Becky's land four years ago, the same time they first came to David's land. It was Becky who called next door and reported it to David. "Our dog ran one of them up a tree!" They both laughed again, but then became serious. "There are some beautiful hundred-plus-year-old oak trees in there," Becky reported. "Big beautiful trees here."

"Do you feel attached to them?" I asked.

"Oh! You bet!"

Becky and Sam are Republicans but often vote independently. "We lean to the conservative side. We're far from liberal," they giggled. They supported the Tea Party in its early stages, but felt that it had "lost its way."

I asked Sam what he thought of environmentalists. He laughed in a friendly way. "I think of what they used to call in the sixties *dirty hippies*. That's all I think about anymore, even though I know for the most part they're educated people trying to do good. You got your fringe group out there, you know, the barefoot, hairy armpit guy." He laughed.

"Are any of your neighbors gung ho for the pipeline?" I asked.

"I don't know anybody in favor of it, except the politicians," Sam said.

"Have you heard about what they're doing up north of Calgary?" Becky interjected. "It's just unbelievable!" With the thought of tar sands soon flowing across her own property just outside the back door, Becky had researched the mining in Alberta. With what she learned of tar sands, she had become worried about leaks and airborne chemicals, as well as loss of financial value to their land. It was constantly on her mind. She said she will feel obligated to tell visitors to her home that the pipeline is nearby, in case something happens.

"And it goes right over the aquifer that feeds our water system; you're drinking it right now," she said. I took a long look at my water glass. "We don't begrudge them doing business to make money, but when you hide information about your product, you lose faith, trust, and believability," she said.

I asked her if she expected resistance. "I expect a lot of people to stand up and resist," Becky said. "We're going to put up notices. We don't have a plan when they come. You don't know what you need to do until you see what they're going to do."

"I think it's more of an individual thing," Sam said. "Because trying to organize people is like trying to herd cats."

Becky stared off across the room, as if thinking what it would be like. "Not knowing when for sure, where for sure, and how for sure they're going to act against your property ..."

Becky was concerned about a mating pair of black panthers on the back part of their land. She had never seen them, but she often saw their footprints. "With all of the commotion with this pipeline coming in, I think they'll leave. And we have deer out here, too—they're *beau*tiful. I'd hate to lose that."

"Well, I won't call that an *environmental* point of view," I said as they both laughed, "but you definitely feel a sense of *stewardship*."

"Oh, yeah! Oh, Lord, yes!" they answered together.

Becky followed up: "I hadn't been here but a month or so when a timber man came out and wanted to clear-cut our whole property. Now, those trees serve a vital purpose on this Earth to *every one* of us. That's not a tree hugger viewpoint—that's just reality. I look out there, and I see nature's beauty."

"It's stewardship," Sam assured me.

"Do you think there's any climate dimension to the tar sands question?" I asked.

Becky was ready for that one. "A lot of people have convinced themselves—with Al Gore's help—that every bit of the warming going on is *strictly* because of things we've done and shouldn't have done. I don't agree with that. Climate change around the world has been cyclical and has been recorded. Five hundred years ago, you've got the mini ice age, and 250 years ago, you have that slow but definite warming. I'm not saying that anything we've done has not accelerated or affected it. What we do has an effect on other things. I don't believe it's the sole cause."

Then she answered her own questions. "Would we be on a warming trend without all this? Yeah, I believe so. Do I believe it is a normal, cyclical cycle? Yes. Do I believe we have contributed some to it? Yes. Are we the sole factor of it? No."

I gave them my usual pitch about the 40 percent increase in atmospheric carbon since the industrial revolution. "If you're going to change the atmosphere that much, you're going to change how it works," I said.

"In those same 150 years, have there been any volcanoes ... or forest fires? Anything independent of man?" Sam asked with an ironic lift in his voice.

I was ready for this one, ever since Curt Carlson tripped me up on it. "There are natural cycles in atmospheric carbon. Some, such as volcanoes and forest fires, put carbon into the air, and others, such as limestone formation and tectonic subduction, take it out. And there's the natural carbon cycle between plants and animals. What's different is that in a very short period of time, we have contributed not just a marginal difference, but a *huge* difference to the atmosphere. The natural balance is out of balance."

"In the media, the human factor is always elevated," Becky observed.

"Yes," I agreed. "Why do you think that is?"

"Because we think too much of ourselves," Sam said.

"Do you think people have a need to ring the alarm?"

"Yes. Why do you think Al Gore did what he did?" Becky was speaking. "Money. Money. Money. We don't think mankind and mankind alone has done all this. We don't think everybody has to go back and live in a tent or a teepee."

"I don't think we *can* go backwards," I agreed. "But we have to go forward in a different direction."

I changed the subject. "What I've been hearing in my own research on the Keystone XL is that this tar sand deposit is so huge that it may increase the carbon level somewhere around 100 ppm, maybe more."

"Something that large would definitely do some global climate change," Sam said thoughtfully.

"And the way that the carbon is going to get to the atmosphere is through the Keystone pipeline. That, I think is dangerous for ..."

"Ev-e-ry-body," Becky finished my sentence.

She thought for a minute about what she had just said. "We just feel powerless to stop what is being driven into our ground. These people come in here and think the laws of the land don't apply to them. I can't see where that should be tolerated," she said.

I TRIED FOR SEVERAL DAYS TO FIND AND INTERVIEW PEOPLE who will be constructing the pipeline. David assured me there are pipeline workers all over the place around Winnsboro, but they are mostly nonunion, and the Keystone is supposed to be a union job. The work had not yet begun, so the workers were not there, but I had been hoping to find someone signed up to work on the Keystone XL when the time comes. I made phone calls and asked around town, but nobody seemed to have a direct connection to the jobs the pipeline would offer.

I did find a union business agent for the Pipeliners Local 798. He was in his office in Tulsa, Oklahoma, so we talked over the phone. The union has seven thousand members around the United States and will be representing those working the Keystone XL throughout Texas. The agent preferred that I not use his name.

"My boss started the project-labor agreement here in Tulsa," he said. "He gave them a reduction in wages and froze prices to do the whole project and all the stations, but they were determined to do the last 112 miles nonunion. There will be around two hundred union jobs, including welders, fitters, and apprentices. Union wages will be about thirty dollars per hour higher than nonunion, including health care, pension plan, and 401(k)s. Union workers will use mechanized welding, where nonunion workers will use stick rods that will take more people and more time. Laying the pipe will take four to five months total. The union workers will come from all over, but largely from within Texas."

TransCanada would do everything it could to make sure the pipeline would be safe, he assured me. But he was worried about the last part of it, the nonunion section. "It's kind of like a chain; you don't take the last link of the chain and weaken it. The last 112 miles will be a weak link. They do not put the money into training that we do. There's a double standard out there. They say it isn't, but it always is."

"Is this pipeline, from your point of view, any different from other crude oil pipelines?" I asked.

"No. ... Maybe a little bit. ... They might be taking a few extra precautions," he said.

"For a lot of people in Texas, this is not a conventional pipeline because of what will be running through it," I told him. "Tar sand is abrasive, corrosive, and pumped hot under high pressure. What have you heard about that?"

"The first Keystone pipeline is in operation right now, and they are refining tar sands oil as we speak. They talk about using lower pressure and adding more valves, but what they say they're going to do and what they end up doing will be two different things.

"The nonunion part of this is going within three miles of my house." As he said this, the pitch and tempo of his voice moved up a notch. "I have friends [whose property] it's going through, and I'll be going out there and looking at it. I'll be looking to see if they're holding them to the same standards as us."

"Are you and your neighbors looking at this as just another conventional pipeline?" I asked.

"They pipeline everybody down here," he said. "They pay them to go across their land, and it's money in their pocket. So they don't mind. But there are people in Texas concerned about it. But all the leaks on the first Keystone pipeline were on sections built by nonunion labor. Later, Keystone [TransCanada] actually hired our local 798 welders to go fix all their mess-ups up there. There were no leaks on any of the union work.

"We preach this to our members: we make more money—we've got health care and pensions—so we have to do higher quality work. We've got a product to sell, and it's got to be a better product." Then he took his voice up another notch. "A vice president of Keystone stood in the union office and said, 'The reason we want to do this nonunion is that we want to take advantage of cheap, abundant labor.' That's what he said."

"Do you or anyone you know working on this have any environmental concerns?"

"I would say the majority of our contractors are 'green contractors'; they deal with the environment every day. They build these things to be environmentally safe. The nonunion contractors don't have the opportunity to work on these big projects; they're not trained. Six years ago, there was a big project here for energy transfer, and they hired a nonunion contractor. They drained wetlands. ... They had no ... remorse for the environment ... There was silt running in all the streams and creeks."

"What is your understanding of the climate situation?" I asked.

"I *do* have an opinion on it," he said. "I do believe there is global warming. There's something going on. I'm fifty-six years old, and I've seen the change. This July is the hottest July on record, and there's a drought everywhere. I don't know how the tar sands is going to affect it. The only thing I know is Keystone was saying they were going to ship this oil to China. They've got less regulation in China than we have here. If it's going to be refined, it's better to have it refined here."

"It's only one atmosphere we're dealing with here."

"I agree on that." He thought for a while, as if trying to summarize his feelings on the project. His voice went back down a notch. "You know, it's going to be built as safe as it can be built. I think it was a good idea to reroute it around the aquifer in Nebraska."

But then he added, "It's not going to bring the price of gas down. They're going to export it out of Port Arthur. The number one export out of this country is gasoline. It's a free trade zone down there where they're going to export it.

"Here's something, too," he went on. "That pipeline is the safest way to transport anything, even if you were transferring groceries through it. But it's got to be built right.

"And you know all these long-term jobs they talk about?" His voice was back in high gear. "These pumping stations have maybe two people each, and there's thirty of them total. Even if you have the maximum of three people on these stations, you're talking only ninety jobs in the whole country. Where are all these jobs?"

July 28, 2012 – Winnsboro, TX
High 96°F – Low 73°F – Precipitation 0.0 inches

An article in yesterday's DeSmogBlog reported that the Pennsylvania Supreme Court declared unconstitutional the pro-fracking bill I mentioned earlier that would have prevented localities from passing zoning against gas wells in residential areas. The court ruled that the bill "violates substantive due process because it does not protect the interests of neighboring property owners from harm, alters the character of neighborhoods, and makes irrational classifications—irrational because it requires municipalities to allow all zones, drilling operations and impoundments, gas compressor stations, storage, and use of explosives in all zoning districts, and applies industrial criteria to restrictions on height of structures, screening and fencing, lighting, and noise."[27] It's been nearly a hundred degrees every day since I arrived in Texas nearly a week ago. There is no relief in sight, and no rain for the foreseeable future.

A LARGE ARMADILLO PASSED THROUGH MY CAMPSITE LATE last night, seemingly on his way to a more important engagement. He paid me little mind, and I returned the favor.

A few hours later, I was awake and participating in the Tar Sands Blockade training session at Susan and Gabriel's farm just outside Winnsboro. About sixty people were there to learn how to construct a nonviolent blockade and use it to delay the pipeline construction. It was so hot that some of the training had to be held indoors.

There was the usual mix of organizations, actions, and agencies. STOP stands for Stop Tar sands Oil Pipelines; it's the group local to Winnsboro that David Daniel founded several years ago. Rising Tide North Texas is a group of students and recent graduates from the University of North Texas in Denton. This was the core group that would be manning and womanning the blockades and tree shelters. They were responsible for organizing the training session. NacSTOP is a group of older folks from south of here in Nacogdoches, Texas, that supports David's work but also has its own list of landowners along the pipeline route. They were more like Mary Pipher's group in Lincoln and my group in Louisville. Trainers had been brought in from Earth First, Greenpeace, and the Rain Forest Action Network. Ron Seifert set up the camp and was coordinating with Susan. The event itself was called the "Tar Sands Blockade," and it would occur the day that TransCanada contract workers stepped on, or tried to step on, Susan's land some time in September. The tree sit at David's would be a sustained extension of the blockade a week or so later.

This training was a little different for me. The blockades to be used would be more than human bodies, and the resistance would be more than peaceably accepting arrest. The emphasis was, as always, on nonviolence, but organizers here were hoping to give police no assistance in removing them from the right-of-way. Lockboxes constructed from four-inch PVC tubing would allow demonstrators to insert their arms into either end and lock them inside using wrist lanyards and carabiners. Plastic barrels to be filled with concrete would also be fitted with lockboxes so that demonstrators would be able to lock themselves to immoveable objects within the right-of-way. Chains and U-bolts around the neck were presented as other possibilities. There was no pressure on anyone to use any of this equipment on the day of the action. Demonstrators were encouraged to find their own level of commitment and take only the risks with which they felt comfortable. Some would use the equipment to delay construction as long as possible; others would stand up immediately when approached by officers and submit to arrest. Some might lock arms without equipment and wait to be pulled apart by police. Many would act in a purely supportive role, avoiding arrest entirely. All levels of involvement would be necessary for a successful action.

These techniques and equipment have been used before with varying degrees of success; they definitely raise the temperature of an event. In keeping police from clearing an area quickly, they create time for the media to get close and opportunity for tactical advantages that can be used in a larger strategic plan. But I, for one, would not feel comfortable directly confronting the physical power of a police force. Force is their province, they know it better than I, they are better at utilizing it than I will ever be, and they will prevail in the end no matter what I do. I was awed by the commitment of many of the young people and a few fogies to risk personal freedom—not to mention life and limb—to protect the earth in this manner. I hoped that they could achieve what they are trying to accomplish; I am not brave enough.

I talked briefly with two of them.

Grace Kagle graduated from the University of North Texas in May, majoring in biology and environmental philosophy. She works on a habitat restoration project. She and about twenty other activists who formed Rising Tide North Texas were working with landowners and other "front line" people, beginning here at Susan's and hopefully expanding to other locations along the route in Texas.

"So you're hoping that this action will lead to similar actions across Texas?" I asked.

"Yes," Grace said. "My particular interest is environmental justice. People of racial minorities and lower income and education are disproportionately affected by environmental issues."

"I've talked to a lot of landowners in other states along the route," I told her, "who want help against the pipeline but are squeamish about an action of this kind on their land."

"They're squeamish about it? Why do you think that is?"

"They don't like the concept of outsiders coming in; they don't like the image of environmentalists. I think that could change, but it depends on what happens here."

Grace Kagle participated in blockade training at the Scotts' farm.

"We're trying to be very careful about our tone, and what we do and don't say."

Grace planned to be at the action on Susan's land, but she also has a trip to South America planned in October and hoped the action would be over before she has to go.

Mike Coleman graduated in the same class. "This is perfect for me, right now," he told me. "I was afraid it would come too soon and interfere with finals. But now I'm out of school, no job, no kids, no family, and I am privileged enough to be able to work on this full-time."

Mike had been working at David's on the tree village. "He does mostly the construction stuff; we do the rope stuff," Mike said. "We have two weeks' worth of food and water, so we will be up there for a while even without ground support. Once we're up there, we'll be able to chill. I have a lot of reading to do."

"Did you do any climbing before this?" I asked.

"No. I started working on it in February. People from Earth First trained us."

When I asked him where his passion came from, he answered, "A deep concern for nature and social justice ... and climate change is very serious—very, very serious. There was a mass extinction that killed 90 percent of life on Earth ..."

"The end-Permian event?"

"That's right. And that one was based on climate change?" he asked.

"As best we know," I said.

"And we're doing it much faster, so that is terrifying," Mike said. "We might kill most of the life on Earth. We could end up in a feedback loop that would be truly horrifying. That's why this action is important. A small number of people can make a big difference."

I WAS UP EARLY THE NEXT MORNING FOR MORE BLOCKADE training. While most were practicing with barrels and lockboxes, I attended a workshop on aerial blockades: tree shelters, tripods, bipods, etc., designed to frustrate police extractors from clearing the right-of-way. Many of these techniques have been used to block logging roads in national forests in efforts to protect old-growth forests from clear-cutting. The trees within the easement on Susan's land are not tall enough for an effective aerial blockade, because extractors in cherry pickers could easily reach the height of any tree sitters. (David's trees are much taller and well beyond the height of most cherry pickers.) Ground-mounted tripods or bipods might be used, but most of the group's climbing resources and manpower would be focused on the sustained blockade at David's.

We were made aware during the morning session that there were several other fossil fuel extraction related events going on around the United States. About five thousand people were attending a "Stop the Frack Attack" event in Washington, D.C.; fifty or so others were marching on a coal mine in West Virginia; and activists in Montana were planning an event in the state capitol against the exporting of strip-mined coal from the Otter Creek region of western Montana. Tweets were flying back and forth between our location and the other three. Bill McKibben addressed the crowd in D.C. and told them about our encampment in Texas. In West Virginia, twenty marchers sat in at the coal mine and were arrested. We cheered them on electronically, and they cheered us on. People were already calling this the Summer of Solidarity.

With a long drive ahead of me, I left the training session in Winnsboro early in the afternoon and headed for home through Arkansas and Tennessee. Approaching the border city of Texarkana, I felt a sudden urge to exit and see if I could find the old bus station in the center of town where, forty-five years before, I had been stranded as a hitchhiker. It was spring break in 1967, my last year of high school, and I was on my own for the first time, hoboing my way across the Deep South. Texarkana was where I ran out of money and out of heart, but that's another story. Right now, I had to get home.

It was 99 degrees in Little Rock, with 108 degrees forecast for the next day. The heat wave was waving on. As I passed through the rice fields of central

Activists practiced linking themselves together with PVC tubes.

Arkansas, I got to thinking about the message the young people of Rising Tide North Texas would be sending out to the world. It was not what I thought it would be. I was a bit troubled. Their action would not be the simple, polite, gentle submission to arrest that I had experienced in Washington, D.C., and envisioned for Texas and Nebraska. There were likely to be scuffles with police—even injuries—and the focus of attention was likely to shift from the climate and the pipeline to how long the pipeline was delayed, or who "won" the action, or who mistreated whom as the conflict unfolded. How would this go over with Susan's neighbors, with other Texans in Houston and Dallas, with Paul Mathews in North Dakota and Randy Thompson in Nebraska? I didn't know. Susan and Gabriel seemed at ease with the lingo and lefty-alternative lifestyle of their longhaired guests, but they had come to know each of them individually. How would these people look on camera? Would they be heroic defenders of the earth or wacko environmental misfits? I feared the latter, but the situation was out of my hands.

The blockade approach seemed more warlike than I was used to. Nonviolence was stressed throughout the training, but this was a more aggressive form of nonviolence. There was greater risk, and a greater sense of *us* versus *them*. Police were not to be spoken to individually, not engaged as people. They were not part of the choreography. Yet demonstrators would depend for their safety on the care and professionalism of the police. The legal risks were greater, too. Locked-in demonstrators—even arm-locked demonstrators—are often charged with resisting arrest, and unsympathetic local authorities sometimes issue felony charges for blockade-style actions. Was this good for young people? I thought of Bill McKibben's hesitation about encouraging young people to begin their careers with a police record.

Yet there was nothing new in this. Isn't this what threatened societies do with their young people? Isn't this the kind of sacrifice young people are asked to make for their country, for their belief, and for their world? If this is war—if the earth is truly in mortal danger—why were we worrying about a few careers and reputations? If, as Mike Coleman feared, the Canadian tar sands will tangle us in an early feedback loop that will culminate in another 90 percent, end-Permian-style mass extinction, what were we doing worrying about a few broken fingers, arrest records, and taser hits? We have to take a stand and stay stood until the world sees. And me, what was I worried about? Getting roughed up by a cop? In my search for a smooth, genteel, relaxed, middle class revolution in understanding of human life on Earth, was I wishing away the hardscrabble realities of the revolutionary process? In my distaste for physical confrontation and for the dehumanization of my opponents, was I not just awfully naïve?

The purpose of a civil disobedience event is *attention*: getting people to tune in. It is a tactic designed to stir human emotion and awareness. But the shaping and direction of awareness, once stirred, rarely comes from a single event. That is a larger strategic question that comes from many events over a period of time. I was, perhaps, too worried about where the attention will go, when there was as yet no attention to go anywhere. I was thinking strategy while my friends in Rising Tide were thinking tactics. Perhaps they would offend a few ranchers and moderate liberals, but they would definitely delay the pipeline. They would get attention. They would be doing exactly what a small number of people can and should do to call immediate awareness to the earth in crisis. They would motivate the next wave—people like me. The next wave will require more people than the first. It will concentrate, hopefully, less on gaining attention than on directing the attention gained. We will point to the situation we on Earth are all in. We will be ordinary people, the "geezer brigade," perhaps, and we will look and dress like ordinary people and aim our message at ordinary people. We will be the mainstream appealing to the mainstream. We will succeed when people on the other end of the media identify with us; when they say, "That is me, out there, defending the earth." That is my dream.

That is when the paradigm will shift.

Alberta

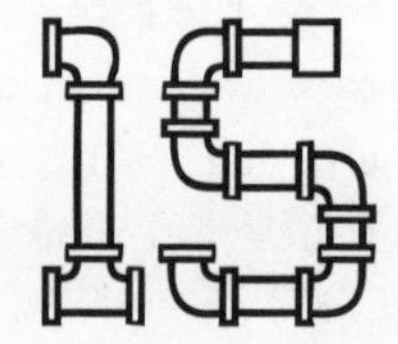

FORT MCMURRAY IS AN ISLAND IN AN OCEAN OF BOREAL FOREST. Houses, hotels, bars, and restaurants spread across a wide spot in the Athabasca River valley and spill over into the neighboring hills, with the forest beginning just beyond the last Burger King or Super 8. There is no agriculture. Except for the river valley, the land is flat, with rivers and lakes scattered on all sides. The tree line is well north of here. There is only one road from the south: the big, four-lane Highway 63, an industrial isthmus that thins down to two lanes a few kilometers north of town, and then to a lane and a half of dirt and mud. There are no roads heading east or west. Not many people pass through Fort Mac; it is not on the way to anywhere else. Only the indigenous Dene settlements lie to the north along Lake Athabasca and Great Slave Lake. To get there, it is best to fly.

Most of the tar sands deposits in Alberta lie hundreds of meters below the surface, too deep to strip mine. They must be drilled and extracted with steam. But just north of Fort Mac, the Athabasca River has slowly eroded the overburden and brought the black, gooey ore nearer the surface. Here, because of surface mining, the boreal ocean has given way to a desert. Open-pit mines, spent sand, and tailings ponds stretch for miles along either side of the road. Double-bodied dump, hopper, logger, and tanker trucks roar through the traffic, interspersed with large, four-wheel-drive pickup trucks with dirt lines spattered up to the door handles. Men in hardhats and reflective vests ride along dusty side roads on their way to man the heavy machinery. Steam and smoke rise along the horizon from upgraders and separating facilities. Small propane cannons in tailings ponds shoot a steady Fourth of July *pop, pop, pop* into the air to keep birds from landing. The wind is brisk, and the air is foul.

On August 4, 2012, this was the site of the third annual Healing Walk, a nine-mile trek "for the healing of Mother Earth and her sacred waters." About

The annual Healing Walk wound around a former strip mine and tailings pond.

140 people took part, mostly indigenous people from around Canada: Sarnia, (Ontario), Yellowknife, Fort Chipewyan, British Columbia, and Fort McMurray. A handful of non-Indians showed up from Edmonton, and there were three or four Americans other than myself. This was not a protest, march, or political rally, but peaceful mourning for the damage done to the land, air, and water by the tar sands industry, and a hope for healing in the future.

The elders began with a tobacco ceremony. We sat in a circle on the ground and passed the pipe, waving the smoke around to our arms and chest, giving thanks for the day and the moment. Waters from all parts of the continent, including Kentucky, were mixed in a bowl and carried the length of the walk. At four points along the way, the four directions were honored with prayers and burning of herbs and tobacco. There were drums and singing throughout. The day began with small thunderheads rolling in from the Rockies and a brief sprinkle of rain, then turned sunny and hot by local standards—about 75 degrees.

The walk encircled four or five square miles of flat, sandy desert—a former strip mine and tailings pond. The water had dried up to a small puddle far off in the middle of the circle. Dump-truck loads of sand lined the edges of the water on all sides. The sand looked fairly clean to me—an off-white color, removed of the tar-like bitumen that gives the ore its name and its carbonated market value. I was warned that, though it looked inviting, the sand was toxic and dangerous. Many people on the walk wore respirators to keep windblown particles out of their lungs. "They don't get all of the hydrocarbons out," I heard people say. "The extraction process releases heavy metals from the ore, and then they process it with more chemicals. Why do you think they have to scare the birds off?" My eyes were not seeing what my mind knew was there. But as we circled downwind of the spent sands, the odor in the air confirmed its impurity, and the sands no longer seemed so clean and harmless.

Three-fourths of the way around, Tantoo, of the local Anzac community, prayed out loud for the waters to flow clean again, the trees to return, and the air to clear. She asked the Creator to bring love and awareness into the hearts of humans, and to help them heal the earth. Nearly all the walkers made it the entire way around. The day ended with a feast, speeches, music, and ceremonial dancing.

BACK AT THE CAMPSITE SOUTH OF TOWN, I SPOKE WITH CHIEF Bill Erasmus of the Dene Nation. He was sitting by a fire at the edge of the woods, fighting off mosquitoes as the long summer day darkened in the sky. I had met him earlier in the day. He beckoned me over with a big smile, put his arm over my shoulder, and insisted I take a photograph. He was well-spoken, with graying hair and ruddy complexion.

"Our language is related to Navajo and Apache," he said, "and our people are settled north of here, downstream along the Athabasca River. Because of the huge amount of water used in developing tar sand oil, our water basin is dropping in the north. We are also concerned because the oil when it is processed uses a large amount of toxic chemicals, including arsenic, that are left in huge tailings ponds. The tailings ponds leach into the environment as the water comes north to us. So we are now finding people with diseases that they never had before. But the water does not stop with us; it flows into the McKenzie River system and the Buford Sea near Prudhoe Bay, and then all over the world. So this development is polluting the world. It is also polluting the air with CO_2. So the Dene Nation has mandated us, the leaders, to oppose any expansion of the tar sands industry. We are directly opposed to the Keystone XL pipeline, which is proposed to go from our province to the state of Texas. We have worked with tribes in the U.S. and with the public in the U.S. We have adopted the Mother Earth Accord with native peoples in the U.S. and Canada, which makes specific reference to the [Ogallala] aquifer. We are also working to oppose another pipeline [the Northern Gateway], which goes

The author (center) with Chief Bill Erasmus and friends.

through western Canada to the coast in British Columbia. The oil is not for Canadian or American usage. It's for the highest bidder.

"Canada doesn't have a plan for sustainable development," he went on. "For us, this means exporting a million barrels of oil a day, which will use up four million barrels of water a day. The other concern we have is that we believe the resource belongs to us. We were never conquered. We were never defeated in war. We have peace and friendship treaties—not with Canada, but with Great Britain. This area we are in is the Treaty Eight area that covers three provinces and one territory."

"How much have your water levels dropped?" I asked.

"We have some of the largest, most pristine water left on Earth," he said. "Where I am from the lake is called Great Slave Lake. It's a hundred miles north to south and two hundred miles east to west. It's like going out in the ocean; you can't see the land. The whole ecosystem is being affected: one, the water level is dropping, and two, it's being polluted. Some people are reporting that water levels are dropping between ten and fifteen feet. Global warming is happening."

"What illnesses are your people experiencing?"

"We're not sure what is happening to our people. But if you talk to people at Fort Chipewyan, which is the first community downstream, they talk about cancers. Many people are dying. They can no longer hunt in that area. The whole food chain is being affected. We are further downstream [in Yellowknife], but we are beginning to feel those effects. What is needed is a plan for sustainable development, for the Obama administration and for the Canadian government to sit down with their people and design that. If we agreed on what the future might look like, then we could have an answer to global warming and to all the changes that are happening to the environment. That's what's lacking. The government here doesn't believe there is an environmental issue, even though the icebergs are melting as we speak. Our people have noticed changes on our land, changes with the water, birds, and animals in the last thirty to forty years, and it has become very evident in the last ten to fifteen. Where we come from, it would commonly be fifty below in the winter, now it might reach thirty-five to forty, and the next day it's minus fifteen, so there are huge fluctuations."

Bill had said something to the crowd at the beginning of the walk that summarized for me the difference between the paradigms I see in conflict over the Keystone XL pipeline. "You said earlier today that 'our purpose in this life is to serve the earth.' That gives meaning to who we are. That puts us in a place larger than ourselves. But it seems that those who are developing the tar sands believe the earth is here to serve us. How does their worldview conflict with yours?"

"Our people believe that everything was provided on this planet, and we were brought here last," Bill responded. "We have laws that are very spiritual. Our job—our goal—is to take care of nature, not to fight it. We can talk about economy, but it has to be balanced. You can't expect to take, take, take from nature and not provide anything in return. There is a huge imbalance; that is why nature is fighting back. There has to be a huge shift. The good thing is that there are many other peoples coming back to that thinking. Our original teachings tell us that regardless of where you come from in the world, other people were taught the same thing. The black people in Africa—they were taught the same thing—the yellow man in Asia, and the same thing with the European people. We were all originally hunters and gatherers, and very close to the land. We were all given the same teachings. We all have clans and families, but people have slowly moved away from that. The industrial revolution has made people more individualistic. It tore people away from the land. But now we are wanting to go back."

"Do you feel that your worldview is spreading to nonindigenous people?"

"Absolutely," Bill said. "With communications, and people traveling around a lot—gatherings like this—we share, we're all human beings, and we find we are closer to each other than we think. But many of us have never talked to each other, and we have to find the bridge."

Bill looked away to the forest and then back to the fire. His expression changed, as if there were something I did not yet understand. "The drum is there to directly communicate with the spirit world," he said. "Whatever you're thinking, whatever you're doing goes directly up to the spirits. The drum is a sacred object that you have to care for; you can't just put it on the floor and have someone step on it. It's like a person, a human being that has to be cared for."

He gestured toward the burning logs. "We have this fire beside us. This fire's alive. It has a spirit. We have to talk to it. We have to care for it. We have to feed it. Our word for fire means home, it means place, and it's comfort—I'm sitting beside it and it's keeping me warm—but at the same time it could be threatening if you don't care for it. It's like everything: you have to take care of it. We have to be very conscious of what's happening around us. We take care of each other, and that's why we have a balance."

FROM BILL'S FIRE, I WANDERED TOWARD THE LAKE, WHICH IS known as Gregoire Lake. The Cree people call it Willow Lake. It is upstream of the tar sands mines, and the water was clear and still. The moon was rising over the opposite shore with a wisp of cloud streaked across its face. A loon called from far offshore.

Henry Basil was sitting with a small group of people by another fire a few feet from the edge of the water. Earlier in the day, I saw him assisting in the tobacco ceremonies. His face was angular, chiseled, aged, and a dark reddish-brown color. I had wanted to take his picture, to have a very Indian-looking portrait in my collection, but that would have shown little of who he was. I resisted. Now I approached him, wondering how to introduce myself, how—or if—he would reveal himself to me.

"I'm not sure what you want," he said, as I stated my intention. Henry was shy and soft-spoken. He was not used to interviews. He had managed a few words to the gathering at the feast after the walk, and I knew there were more within him. But I hesitated. People were talking on all sides, and someone was beating a drum behind us. We sat in silence, for a minute or two, looking at the moon over the lake. He was nervous at first, but he warmed as we spoke.

"I was impressed by your story," I began, "When you were taken from your home at age five ..."

"I remember we were coming back from our hunting grounds this time of year, after hunting caribou, to have meat for the winter," Henry began. "This big boat—a freighter—came to the shore."

"On Slave Lake?"

"On the east arm of Great Slave Lake. There were five or six families with children. We had been along the tree line, where we hunt caribou. This man dressed in black with a cross, he's the one who grabbed me and dragged me down to the boat."

"You had no warning?"

"No, none," he said. "My mom was coming down after me; I was hanging on to her dress. No one could do anything. They put us in the bow of the boat, seven children. It was dark in there. We traveled in the boat somewhere, I don't know where. I don't remember much. When you are traumatized, you lose touch with reality. We came to a convent. There was a buggy on the shore. I don't remember if it had a horse or an ox. Sometimes ... I don't really like to talk about it ..." He paused, looking down at the fire, his mouth hanging. "But it heals me." He looked up. "The next thing I knew, they took us into this building. There was a woman dressed in gray, with a black hood, wearing a cross. They took our clothes away. I remember being in a blanket. They kept us there for four years. I don't remember going to school; I don't remember sitting in a class. All I remember was the church. It was hard."

"You didn't see your family those four years?"

"No.

"After the fourth year," he continued, "someone who had been there before explained to me, 'This is what's going on.' After that I started remembering things. The way they treated me up there was horrendous. It was horrible. They controlled us through hunger—we never had enough to eat. The food was horrible: fish, rotten fish. They make you eat that. If you don't eat it, they save it for your next meal. That is the way they treated me. You don't know what is happening. You do as you're told and that's it. They used to call me '*sauvage*.' I didn't know what it meant. I saw a friend one time and said to him, 'I have a new name, Sauvage!'"

"That's who you thought you were!"

"Yeah! And the other boys, they were saying the same thing. 'Hey, I got a new name!' We didn't know what was happening."

For many years after that, Henry rambled around Edmonton. "Lying, cheating, and stealing—that's all I knew. I didn't know what else to do," he said. "But I have been on a healing journey since 1974. I met this elder in Edmonton. I was crossing the street. I had a really bad hangover. There was a bar across the street. Someone behind me said, 'I used to be just like you, son.' I didn't know where the voice came from, and then he said it again. The first thing in my mind was that this could be an easy target for me. But instead of giving me money, he

Henry Basil was taken from his family near Great Slave Lake at age five.

took me into a café next to the bar and bought me a meal. We talked for quite some time. I was tired and hungover, and I was ready to ask him for money, but he took me to his house and said he would take me to a treatment center the next day. At the house, his wife washed my clothes, and I went upstairs to sleep. When I woke up, there was a chair next to my bed with all my clothes all nicely folded and clean. My socks were all clean. The elder knocked on my door and said, 'Breakfast is ready.' That's when my healing journey started."

"Out of his love," I said.

"Yes. I was thinking it was an angel. He took me to the treatment center. I never saw him again."

That was only the beginning of the journey. Henry "fell off" and started drinking again. He was sober for six or eight months at a time, but he fell off several more times.

"This went on for some time. I spent a lot of time in drunk tanks. No one really could control me. I thought of suicide many times. I was on the streets five years. After I sobered up the third time, I came home to Yellowknife and got into my traditional values: sweat lodges, hunting ..."

"So you found out who you were," I said. "You weren't *Sauvage*."

"Yup. I was somebody, *somebody*. And the time went on, and I fell off again. My dad died, and I didn't even go to his funeral—that's how miserable I was.

But my mom, she helped me. She told me stories of the old times. I started hunting, fishing, and trapping again with teachings from my mom and the elders."

"The angel is inside you. You are helping others now."

"In my prayers, I always mention Angel. My mom taught me that."

"Do you feel that coming here to the Healing Walk is part of your trying to help others?"

"Yes, oh yes! Yes," Henry said. "Since I started seeing what was happening to my people and the lands, I became more outspoken, did translations, went to meetings. I started working for archaeology. We found artifacts, graveyards thousands of years old. One artifact was carbon-dated back almost 30,000 years. I offered tobacco at the sites. One time I was walking up a river and there was a little clearing before you hit the lake. I noticed a big pile of pebbles on the left-hand side. They were round and smooth; some were black and some quartz. I didn't tell the archaeologist about it. But when I went back home, I told my mom. She said, 'Yes, those are offerings. Every time we would go to the barren lands to hunt we would leave a pebble there.' I showed it to the archaeologist the next year, after I knew from my people what it was."

More people were gathering around the fire where we sat. The drum was louder and closer, and it became hard to hear. Henry was talking about the spirits and dreams that came from his work with the archaeologists. "I became a runner," he said. "I was working at the fishing lodge and had to be there at six in the morning. I would wake up at four and run twelve miles before work. I ran four marathons. My fastest time was two hours, fifteen minutes, and seven seconds."

"Two fifteen? That's Olympic scale."

"Yup."

"That helped you from falling off again."

"Yup. And my mother helped me."

The drums were at the fire now, and people were singing. Henry wanted to keep talking, though I could barely hear what he had to say. I was trying to steer the conversation toward the tar sands, the pipeline, the climate, the earth. "What do you think we have to do now for Mother Earth?" I asked. "By *we*, I mean your people, my people. I'm American, a white man ...."

Henry's answer was a lot simpler than what I was looking for: "Do you remember I talked about *lie, cheat,* and *steal*? That was all I knew. Well, I rephrased that to *love, kindness,* and *compassion*. It took me a long time to be what I am now. I spoke to other peoples in other nations, I did sweats," he was saying over the music. "That is my story. When the elder found me, he brought my world back to me ... I didn't know what it was."

There was something I still wanted to know: "What happened when you first noticed that the water in the lake was dropping?"

"When the water level dropped," Henry said, "it broke my heart deep down inside. I am an emotional person. I thought, 'What about the fish? What about the birds that use this water? Where are they going to go?' The fish no longer come to the places they used to come to. This is my home, Great Slave Lake. This is where I live."

August 5, 2012 – Fort McMurray, AB
High 74°F – Low 48°F – Precipitation 0.0 inches

"The pipeline," I learned here, does not mean the TransCanada Keystone XL to Texas; it means the Enbridge Northern Gateway pipeline across the Canadian Rockies from the Alberta tar sands to a deepwater port along the coast of British Columbia. Enbridge is the company that brought us the recent spills in Michigan and Wisconsin. The Canadian government is hell-bent on getting the tar sands (or oil sands, as they are more politely called by their promoters) from the landlocked center of the continent to its edge at any cost. The Northern Gateway pipeline is opposed by First Nations groups, many of whom were represented at the Healing Walk. Stewart Phillip, grand chief of the Union of British Columbia Indian Chiefs told the crowd he is worried about supertankers plying the island-strewn coastal waterways where his people hunt, fish, and live. Prince William Sound is still suffering ecological damage from the Exxon-Valdez oil spill twenty-three years ago, he pointed out. The traditional ways of indigenous peoples could be wiped out by a single accident of similar scale, and the tankers are much larger now. His people are being asked to take that risk for the benefit of the tar sands industry. B.C. Premier Christy Clark also opposes the pipeline. But if it is built, she will expect a "fair share" of oil revenues for her province.

But Enbridge may have another card to play. Bob McLeod, premier of the Northwest Territories, has offered a northern route to the sea through his domain, in exchange, no doubt, for a slice of the pie. Like a slowly filling basin, the Canadian tar sands are cresting the rim of their geographic confinement, searching south, east, west, or north to break passage through to the sea. Our appetite is its gravity.

"I WAS IMPRESSED BY YOUR PRAYER YESTERDAY," I SAID AFTER introducing myself.

Tantoo Cardinal was folding her tent. She was about my age, a Cree, and had recently moved back to the Anzac community (the first nation community in the Fort McMurray area) after living on the West Coast, where she was involved in acting and filmmaking.

"My worldview was incubated here in this community, Anzac," she began. "Years ago we didn't have a road. There was just a train that came through one day of the week. Two days later, it would go back to Edmonton. That kind of situation builds a close-knit community. I remember going to a campsite far from home when I was a child. We were berry picking, and Mom made a streak into the bush and pulled a cup off a tree. People would leave things for other people; there was no sense of ownership. The same thing with cabins in the bush. You were welcome to go in and use whatever is there; it is meant for that. It's a community that is intertwined. I felt a breech between people at an early age when I went to the city and saw the disdain for my people."

"In Edmonton?" I asked.

"Yes; a lack of respect. It hurt, because the people who raised me were all kindness and compassion. It was the *way,* the way. You just knew intrinsically that the earth is alive: everything is alive."

"It wasn't like a realization ... "

"Oh, no," she said. "It's like breathing, and if you take anything from the bush, you leave something. Tobacco was a sacrament. You leave it as a gift. You appreciate. And you recognize the bounty of this living earth. And I was thinking, 'I wonder what kind of sacrifice Syncrude and Suncor put down!' " She laughed. "And judging by the effects it's having on people, if they did put something down, they better put something else down, because it's not in balance. They haven't given anything back that's of any value. There's been no gratitude. When the tar sand deposit was discovered it wasn't, 'Oh! Now we can do this for the people; now we can do this for the earth!' There was no breath of gratitude. Indigenous people are still connected to the earth, despite the attempt to sever our relationship to the Creator. The thread was pretty thin when we started coming back, but that is what has sustained us through the barrage, the genocide, the holocausts."

"Do you feel people are coming back now?"

"Yup," she said. "Communities have been healing. We're taking back our culture, taking back the Creator, taking back fundamental relationship understandings."

"The worldview you are expressing now: you were born into that. I was not born with it ..."

"That's right. My father's people didn't discover that the earth was alive until the sixties. We've been trashed. They considered us primitive ... not quite as

Tantoo Cardinal was an actress and filmmaker before moving back to Alberta.

intelligent … not capable of huge concepts. How would *we* have any answers to anything?"

"I see a lot of people in mainstream society in America and in Canada that are looking for something," I said. "They're not sure what they are looking for, but they need help."

"They do," she said. "And unfortunately they don't look in the right places. But … now, maybe they will. With climate change, they finally realize it affects all of us, not just Indians. With the tar sands, it was just Indians. It doesn't really matter; they die easy." She laughed, and looked me in the eye, knowingly.

"It's the climate issue that has brought me here from Kentucky," I said. "You have the tar sands here; we have mountaintop removal. There's fracking everywhere. But the issue that ties it all together is the climate. That's why I decided to seize on this issue."

"Imagine reaching in with your dirty hands and grabbing a piece of liver out of the human body, or a hunk of pancreas," she said. "That's what it means that the earth is alive. This is all a part of her body. It's a cancer. The earth has cancer. People think they can keep doing it. They can't be that stupid."

"No, they're not stupid. They're working within a worldview that is limited by money, and they make their decisions based on profit."

"And they made sure we didn't get our hands on the money. They handled us for generations. That was a really smart tactic on their part, to tear the children away from their families and leave the parents with no children. And to program the children that their parents' ways are the ways of the devil—teaching them to fear it. Ripping their language away. The Bible is not based on natural force; that was taken out of it. There are some pieces missing to the Bible, and not only that, it was written by men. It was distorted for the purposes of their own power. Nature doesn't lie to you."

I felt the strength of the living world in her words, but a weakness in how they were heard by others. "Now that society is going to have to reconfigure itself," I asked, "how are we going to live on the earth, how are we going to get human activity within our understanding of the whole earth system? How are we going to do that?"

"They've got to recognize natural force, and that means solar energy, wind energy," she said. "That's the only thing that's replenishing. A Hopi elder shared with us one time a vision they had gotten in their spirit councils. He said, 'Today, people have every confidence in the cycle of condensation—of clouds and rain, etc.—every confidence in it. But if things keep going as they are, the earth will suffer a stroke. When that happens, there is no cycle.'

"Years ago, when Earth awareness was rising in the larger society," she went on, "I was very excited that they were using some of our quotes from our chiefs and some of our understandings were moving into the environmental movement. It was a key time. Then, Reagan came along and said, 'If you've seen one red cedar, you've seen them all,' and the whole structure just went in the other direction. So we have poison all over the earth now. We weren't incorporated in to be in the front lines. Our drum wasn't brought in; our songs weren't brought in; our ways weren't brought in. We didn't have that respect."

"Even from the environmental movement."

"Yes. I will always remember that sensation when I heard Reagan on the radio. There was a massive chill in my body. Society just moved completely away from the earth. I felt the world had gotten away from us, but coming back home here there was hope yet, there's *hope* yet. We haven't destroyed it. Our air is not like in California. But nobody was listening, even among my own people. They were elbowing me out of the way, 'Don't talk like that, we need the money,' you know? So we're all being educated," she said.

"I think with climate change we're all beginning to experience it. We're in a better listening mood now," I said.

"Well, you keep working on that. Open up their ears."

MELINA LABOUCAN MASSIMO WORKS FOR GREENPEACE CANADA. It was her outreach to the people of Appalachia that resulted in my coming here to Alberta. She was a key organizer of the Healing Walk. Her community of about five hundred Lubacon Cree is from the Peace River tar sands region, several hundred miles southwest of Fort McMurray.

"Before my dad's generation, people lived off the land," Melina told me. "Horse and wagon. With his generation, the paved roads came. His parents hid him when the residential school people came, but he has a master's now in linguistics and education. His job title is Aboriginal and Cree Curriculum Development. A lot of people used to live by subsistence, but because they are being pushed off the land, or the land is leased out, they cannot live the way they used to. They can't pick berries or gather medicines. Gardens used to be family based, but now agriculture is more of a business."

"So, when did you become aware of the destruction by the tar sands industry?" I asked.

"People didn't really talk about the tar sands *per se*, even up to five or ten years ago," she said. "It was always called 'oil.' There was a lot of conventional oil, so it became normalized. Just oil. But there hasn't really been a discussion of the difference between oil and tar sands, which is an unconventional type of fossil fuel. No one ever thought of it that way. It wasn't until my master's work that I started researching tar sands. I saw that the deposits lay exactly where my community is, and where we are right now, and I saw how big the project is. It's one of the biggest projects in the world. It's a huge project area. So I came to realize that this oil and gas lifestyle is going to get worse." Melina does not always like living in Alberta because it is so resource-driven. But she has stayed and focused her work on environmental issues. "As a native person, I've always been about the land, about our abilities to practice our livelihood and our traditions on the land. It's a much bigger issue than I realized at the time; that's why I started working on it full-time."

"So," I asked, "where does your passion for the land come from?"

"I grew up in a northern, rural community with no running water, no infrastructure, but my mom moved us into the city for a better education. It was a big culture shock. Like, this is totally not what I expected. I was shocked. There were so many more resources, libraries, books. But growing up with grandparents that lived off the land—fishing, trapping, hunting—and going out on the trap line with them or in the horse and wagon, just seeing how beautiful and diverse the boreal forest is, and the ecosystems and all the life that was there and how it provided for our families with clothes, beading and moccasins, medicines, and food. That's what I grew up with. Every time we would go visit my grandmother, she would cook up moose meat and give us *banik* [deep-fried bread]. People would always give each other meat or berries. I learned from my dad how to harvest certain medicines. You know that this Earth sustains us and gives us life. It makes you want to protect it all that much more. This isn't just a piece of real estate. Developers who are not from here don't have that connection. They think it's just some plot of land, but it's our territory, our homeland, where our ancestors are buried. It's not just us standing in the way of 'progress,' it's us caring about the very foundation of how humans can even live in a good, healthy way on the land."

"That seems to come naturally," I said.

"It's a cultural thing. You kill the land; you kill us. You kill the land; you kill our culture; you kill our livelihood. It's not like you can kill this plot of land and we will just move to the city and become consumers.

"When I came to Louisville," Melina went on. "I don't remember if I said anything to the crowd about it, but when I walked in and saw this picture of a mountaintop removal site, I remember thinking, 'That looks exactly like the tar sands.' Then hearing everyone's stories about the contamination, deaths in the families, etc.—it's the same thing."

Melina is continually frustrated by the tendency to pit the environment against the economy. "The economy doesn't keep us alive; it keeps us *functioning*. The environment is what keeps us alive. I just wish there was a way we could be transitioning off fossil fuels to see what a world would look like with solar and wind. I think it's the mind frame that is really different."

"We may have to develop alternative energies, but why not? That makes jobs, too!" I said.

"Right! Exactly. It's all about domination and control of the masses. The oil companies have their infrastructure set up, and they don't want people to stop buying their products. And the people getting a wage from it are really participating in their own demise. I guess that's coming from a different worldview."

"It's subconscious," I added.

"It's cultural teachings that give you a different understanding of the world, how to acknowledge what's important in the world. Being out on the land makes you appreciate what it is," she said.

"Is that what's living through you in your work now, your experience of being close to the land as a young person?"

"Yes. Those are some of the best memories I have. I remember going on the horse and wagon with my *cohcum* and my *mosum*." Melina had a far-off look on her face. "They would go out into the bush for a month or two at a time. There were summer and winter camps. It was seeing that from a young person's

Melina Laboucan Massimo works for Greenpeace Canada.

eyes—the forest is so *vast*. I thought it was never-ending, all-encompassing. It was *so* dark at night, and *so* quiet."

"That is the world you live in."

August 6, 2012 – Fort McMurray, AB
High 76°F – Low 44°F – Precipitation 0.0 inches

According to today's Nation of Change, the dry, hot summer is causing water temperatures to approach 100 degrees, killing off thousands of fish in streams and rivers in the United States. About forty thousand shovelnose sturgeon were killed in Iowa last week as the water temperature rose to 97 degrees. Sturgeon, catfish, carp, and many other species of fish in the lower Platte River are boiling in the drought-stricken heated waterways.[28]

"It's something I've never seen in my career, and I've been here for more than seventeen years," said Mark Flammang, a fisheries biologist with the Iowa Department of Natural Resources quoted by the Associated Press. "I think what we're mainly dealing with here are the extremely low flows and this unparalleled heat."[29]

THE LIVING BEING OF THE EARTH IS NOT AN ABSTRACTION FOR the indigenous peoples of Northern Canada, nor is it, I believe, for indigenous peoples anywhere. It is not a realization from scientific studies of ecosystems and weather, not a Gaia hypothesis. It comes from culture, from early childhood teachings, and from direct exposure to plants and animals, rivers and streams, lakes, forests, and storms. But this understanding of life does not fit modern society, and indigenous peoples everywhere have found it squeezed out of them through systematic programs of assimilation.

But it has come back. The Cree and Dene people I have spoken to are no longer embarrassed by who they are. They talk openly of spirits, of Mother Earth, of the living forces of air and water, and they know that their words to each other are understood as spoken. They feel that their ways speak to humanity as a whole, although they also know that few outside of their cultures are hearing what they say.

The reality of a living earth is not present in the mainstream culture. One cannot speak of it without apology in normal social relations. At the office, or with friends, we do not speak of water, air, or earth in spiritual terms, even if we know them as such, because we cannot be sure what is heard on the other end of the conversation. Spirit is fancy, imagination, superstition; it is not accepted as experience. For most people, it is pretend. If explained, it is not understood. When we hear of the Mother Earth Accord, we think, "Sure, they can talk that way. They're Indians." But we cannot talk to each other that way.

Science has come to understand life in terms of complexity. There are so many organic chemicals seeping through membranes, so many genes directing new growth, so many electrons flying between neurons, that life arises, somehow, from the traffic. We know more and more about how life works, but we do not know what it is, or where it comes from. I do not believe we ever will know what life is through science or through the perceptual experience on which science is based. I do not believe we will ever see, feel, smell, taste, or hear life itself. Even in interpersonal relations, we see how people move, hear what they say, and watch how they react to pleasure or pain, but we do not hear or touch life in them directly; we acknowledge it in them only through assumption. Awareness of life in others is, therefore, a leap of spirit. The order people create in what they do and the symbolic impact of the sounds they make with their mouths lead us to believe they see and feel as we do, but we do not see or feel it directly. We do not perceive life as an object. We know that our friends and family are alive only because our hearts are open to them.

The trees do not speak in words, and the lakes and streams do not flow in the language of English or Cree, but the earth moves always toward higher creation and new birth. The birds and fish create order as they hunt and build nests. The forest grows back where trees have fallen. This is what Henry Basil saw along the shores of his homeland after years of street life in Edmonton, and what Melina's grandparents showed her at Peace River. We cannot see or touch directly the subtle life of the earth any more than that of friends and family, but it is there. The earth is alive. It is bigger than science or language.

Symbiosis begins by seeing life in those on whom you depend.

I FOUND ADAM, STEVE, AND BRADLEY AT A BAR IN THICKWOOD, just up the hill from the center of Fort McMurray.

"I'm up here because my daughter needs money to go to college," Steve said. He was reluctant to talk at first, but once he got started, there was no stopping him. "I wound up working pumps for an oil sands contractor. It's a different place from [Vancouver] Island. I've been to college twice, got my operator's [license] in heavy machinery."

"What kind of machinery do you operate?" I asked.

"Excavators, dozers, backhoes, loaders, forklifts. I was in the lumber industry for 25 years in [British Columbia], but the money is way better here."

"They take you away from home, so they have to compensate you for it."

"Yeah, they do take you away from home. Home is like winning the lottery," Steve said. "There is no place like home." The waitress came by at that point and he asked her, "How good is home for you?"

She was also from B.C. "I don't get to go home very much," she said. "I've just been here two months—I have to stay at it."

"What brought you up here?" Steve was doing the interviewing for me.

"Money," she said without hesitation.

"You only find good attitude out of B.C. people," Steve observed.

"Did you know each other back home?" I was asking the table as a whole, but Steve was still talking.

"No. Never seen each other in our life. But we're like family. Only the strong survive."

"Days up here mean nothing," Bradley added. "It could be a Monday; it could be a Friday. The only day that counts is your day off. Even Christmas Day. It doesn't matter."

"Isn't the cold weather around that time of year hard on machinery?" I asked.

"I'll tell you what's hard on machinery ..." Steve began.

"Yeah," Adam laughed. "Brad here." We all laughed.

"A lot of Newfies here, too?" I asked.

"Yeah, there are Newfies here, apparently," Steve answered. "I've been trying to go to a Newfie bar, because they say it's a good atmosphere, a lot of good people. I have yet to set foot in one."

"Do they hang out together?" I asked.

"Yeah, there are a lot of groups that stick together. But, being a multicultural place, I find it really awesome, except for the fact that people up here have bad attitudes."

"Because their heart is somewhere else?"

"Like anywhere else, we put in our time, put in our hours, get our paychecks."

Steve tries to get home once a month, but it is usually every two months or so. But he gets stir crazy and doesn't stay long. He sees his daughter for a day or two, then his family for another couple of days. He is always thinking of home at work, but he wants to get back to work while home.

"We work twelve-hour shifts," he said. "But they pick us up an hour and a half before we're due at work; it takes that long to drive up there. So it's really a fifteen-hour day. We do five, five, and five: five days, five nights, and five off, but that never happens. I suck in as much as I can until I'm brain dead. Then they make me take time off. I've been up here for seven months. There's a lot of turnover. Some people come for two, three months, but they can't take it. Adam here is a newbie ..."

"How long you been here?" I asked Adam.

"Two months. I have to pay for one more semester of school. It costs me about ten thousand dollars a semester."

"Where you going to school?"

"South Alabama. I went there on a baseball scholarship, but I got hurt and they took away my scholarship, so now I have to pay. You pay the state rate, then you pay the out-of-state rate, then they double it for international. Then there are living expenses."

"What is your work here?" I asked.

"We pretty much pump fresh water into the plant and onto the wastewater so it has a chance to settle," Adam said.

"So here we are," Steve piped back in, resetting the stage. "Making money. Trying to do the best we can. I'm putting aside money to build my own house. The oil industry is here to stay. We all count the days till we get back. We work hard, but we count the days. It's long hours; it's a struggle to stay awake. We all come here for a reason. You don't know anybody personally the first little while, but then they become family, and a good family, too. We all got each other's back. We help each other out."

"I didn't know nobody when I came up here," he continued. "I was like a stranger in a strange land. They just toss you up here, in this house. I didn't meet anyone. Either you make it or break it. That's all there is to it."

Some of the heavy equipment used at the mines near Fort McMurray.

"Does the company own the house?" I asked.

"Yeah," Adam said. "That's kinda good, because they put all the same guys together. If somebody sleeps late or something, we can wake him up. They drive us to work in a van."

"Do you have a car here, too?"

"No. We live nearby; everything's right around here. We walk."

"Is it easy to save money here?" I asked. "There's drink, girls, drugs ..."

"Drugs are for duds," Steve said. "It's easy for Bradley, here, to save money, because he never buys us rounds ..." Everyone laughed. Bradley is the tightwad in the family.

"It's easy," Bradley confirmed. "Don't buy a big truck; don't pay fifteen hundred a month on rent like everyone else around here. You're still making six, eight grand a month ..."

"And you're packing it away."

"*Exactly.*" Bradley reminded me of myself at an earlier age. I was working in New York and squirreling money away to buy the farm in Kentucky.

"If you have no education, you can come up here and make what doctors and lawyers make," Bradley said.

"Do you feel like this is home, though?"

"I could stay here for a while, maybe seven, ten years. Then go back home and live there for ten, then back here for twenty, then back for ten ... But I can make a down payment on a house from what I earn here. I need a hundred thousand dollars, so I can *do* something."

"What if the price of crude oil drops?"

"There's no worries. I've been reading this book: *Tar Sands*."

"Nikiforuk?"

"Yeah! Who are ... Wait ...Are you? .... What's your name?" Bradley remembered I was an author.

"Avery ... not Nikiforuk."

He relaxed. "I've been reading it at work. He, like, knows more than anybody about it. But he's kind of like *mad* at it all. He's very negative. He talks about the pollution, how they ignore the pollution, little particles in the air."

"What do you think about it?" I asked.

"Those stacks ... they just puke out the white fly ash," Bradley said. "Thing is, nobody can put a finger on how bad it really is. You know, I come home feeling a little grogged down. But the guys who work on the loaders and the crushers, it's ten times worse. After a few years ... I've seen guys who look so old and beat up. Nobody can really pinpoint what it is. Nobody really wants to talk about that." But Bradley is not worried about running out of work. "This all got started thirty, forty years ago, but only got going big in the last decade. We're at the very start of a thing that's going to go on for five hundred years."

"Do you know anything about the climate impact of burning all this tar sand?"

"CO_2?" Adam asked. "Ninety-eight percent of it is from volcanoes. I don't think any one of these sites up here can be blamed single-handedly for CO_2. I don't like to argue about it, because nobody's ever going to be able to shut it down."

August 8, 2012 – Fort McMurray, AB
High 82°F – Low 51°F – Precipitation 0.0 inches

At the Oil Sands Discovery Centre in Fort Mac, they show you what to discover and where to discover it. I was expecting a barrage of carbonated PR, self-promotion, and heroism—searching to meet our ever-growing energy needs, etc.—and I got it, but there was quite a bit of useful environmental information as well. The presentations and exhibits were more balanced than I had thought they would be. The total reserve in Alberta is 1.7 to 2.5 trillion barrels, 300 billion of which is extractable with current technology. A sign at the exhibit called it "The Biggest Known Oil Reserve in the World," but the guide who showed us how oil sand is processed and distilled said the Alberta reserves were actually third in the world, behind Venezuela and Saudi Arabia.

One exhibit demonstrated how the in situ, or steam extraction process, damages the environment without destroying the entire boreal surface. The first step is dissection, a linear slicing of the forest by cut lines for roads, power lines, and pipelines. This creates discontinuous habitat and allows penetration of the forest by invasive species. Perforation happens with inroads into the forest from the cut lines for pump stations, equipment, housing, supply yards, etc., which destroys a greater area of the boreal habitat. Fragmentation begins when perforation inroads meet one another, leaving islands of isolated forest. This type of environmental degradation is common in outlying areas beyond the surface mines, and in the Peace River and Coal River deposits, where 80 percent of the tar sands are extracted. The exhibits suggested ways that the damage could be minimized.

Toxicity was not discussed in any depth, and the climate impact of trillions of barrels of new hydrocarbons was not mentioned. There was no attempt to hide, only to normalize. There were exhibits where visitors could use fixed tools to scoop and stir tar sands without, of course, touching them.

FROM THE DISCOVERY CENTRE, I BOARDED A BUS FOR A TOUR OF the Suncor facility. We drove north along the Athabasca River. At the main gate, security people boarded the bus, walked up and down the aisle, looked us over, smiled, and waved us through. The largest trucks in the world were busy dumping ore into a crusher that fed conveyor belts streaming into separating facilities. Each truck, fully loaded, weighs 400 tons—the weight of a jumbo jet. The Komatsu trucks cost more than $4 million each, the Caterpillars $6 million. They are delivered in pieces and assembled on location (after trying to drive them overland through Montana failed, as Ron Seifert related). Drivers have to climb ladders twenty-one feet tall to get into the cab. The tires are around eleven feet in diameter and cost $70,000 each. A tire change for a six-wheel truck costs more than $400,000 and lasts only nine to twelve months. Oil sand is an unforgiving substance to work in, on, or around.

Above the river, we viewed a reclamation site where Suncor had planted several hundred thousand trees. The trees were still quite small, and the site looked more like a field than a forest. But they no longer plant trees in rows, and some wildlife had begun to move in. As we stood outside the bus looking over the reclaimed acreage, a puff of wind blew a strong rotten-egg smell our way. When someone asked the tour guide about it, she pointed to the

The largest trucks in the world were busy dumping ore into a crusher.

sulfur-extracting facility across the road. She offered no apology. One of the tourists said, "We were told the smell was going to be terrible up here; but this isn't so bad."

From the Suncor plant, we headed farther north up Highway 63 to where the Healing Walk had begun six days earlier. We followed the same route, circling the same former mine in the opposite direction. We saw the same water, the same tailings, the same off-white sand piled in mounds around the edges of the water. I was looking through the lightly tinted glass of the bus windows at the same thing I had seen earlier. *Tar sands* were now *oil sands*; the black had been bleached white; toxicity was nowhere to be seen. "There is less in this sand," the tour guide assured us, "than when we mined it. The bitumen has been removed." Square miles of desert whiteness spread before us, like an ocean beach suspended in the midst of the boreal forest. It was all so normal, so plain, so matter of fact. This is simply what we have to do for energy.

Birds circled high overheard, considering a waterborne landing. The popping of propane cannons barely penetrated the whirr of the bus engine.

BACK IN TOWN, I MET JIM AT THE BAR IN A PIZZA JOINT. IT WAS mid-afternoon, and the room was nearly empty. He had a small, narrow build and wore a ragged baseball cap. I judged him to be around thirty. He was skeptical when I introduced myself.

"Why are you writing this book?" he asked. "Is it going to be on the pros and cons, or is it going to be only the pros, or only the cons?"

"It's a big issue that opens a lot of questions," I answered. "Questions about jobs, economy, environment, landowner rights, the climate. I'm trying to find out what people's stories are. I'm saying what they say."

"I don't have a lot to say on this ..." Jim went back to his beer, and I thought I had lost him. But then he looked directly at me. "I love the environment. I live in Wooster, B.C., one of the most beautiful places in Canada, where the Olympics were held. If I ever heard of a pipeline going through there, I'd be disgusted. But the fact of the matter is, the way the world is going now, we need energy, and fuel, the fossil fuels ... we got to get it somewhere. It's as simple as that. I hate the job. I hate hearing about the spills, and everyone bitches about it. But it happens. That's too bad. But if none of these pipelines weren't there, well, guess what? We'd be f-cked. We could find other ways, but at this time they're not feasible."

Like most people I met, Jim opened up as he spoke. "We're just simple people. Like my old man—he's been a pipeliner his whole life; it's not his fault that all of a sudden there's a crack and a spill. There's inspections; there's X-rays done on every single weld, but then there's natural phenomena. I don't know the whole thing."

"Are you a pipefitter?" I asked.

"I'm learning, I'm an apprentice. And you know, there's the Keystone pipeline—all my friends are against it. But this is the line of work we're in. It's not our decision. The government wants this; it's all about money. Money rules governments. It's as simple as that."

I bought him another beer.

Jim wanted to keep talking, but I could tell he was depressed. He often interrupted himself, or me, and then took back what he had just said. He had just found out a few hours before that his thirty-three-year-old brother in Ontario had been diagnosed with lung cancer. Not from smoking, he didn't think, but from working in a poorly ventilated welding shop. "My dad is ready to take out a loan—he doesn't care how big it is—to pay for treatment in the States. So what do *I* do? I'm f-cking alone here right now. I don't have anyone," he said.

"How long have you been here?"

"Since January. I did a pipeline job. That was good. That was fun."

"Have you been back to B.C.?"

"Once. B.C. is the best place in the world to me."

I was drinking a Molson M and ordered another. Jim stopped me. "What are you drinking that stuff for? Get a Molson Canadian."

"Okay ... talked me into it. We Americans can't tell the difference, you know." We laughed. "Tell me more about your work."

"I work out in the mines. I do a good job—a great job. I always do a good job. I do ten-hour days, ten days on, four days off. We're welding pipe together for these giant ponds, the tailings. I don't know how it works. They're trying to recycle as much water as possible to use less natural water. It's actually pretty cool."

"That's a good thing. That way they don't use as much water from the river."

"It's a damn good thing," he said.

"What do people do on their days off?"

"Drink." The waitress interrupted, filling a beer mug at the bar for another customer.

"A lot of people get involved in drugs," Jim went on. "I don't do the drugs. Anyway: why am I in the business? Because there is money to be made. I don't control decisions. Money rules the day. That's the way these big companies work." He was becoming more cynical with the beer, and with his brother's illness on his mind. "To me, I hate it. The paychecks are good, but I hate this place. I live in camp—I call it jail. There's security everywhere."

"What are they worried about with security?"

"Oh, I don't know, probably drugs and alcohol," Jim said. "My dad tried to drop me off one time—we had dinner together—and I showed them my badges at the main gate, and he's like 'I'm just dropping off my *son*,' but they were like 'Oh, we can't let you through.' Suncor camp: it's beautiful. Love it to death."

Jim took a long draw on his beer and looked around the room. "You know," he said in summary, "this is generally your oil. If they don't like it, you guys will just take over our country and that's it. All of a sudden Canada is America. Although you'll have a lot of hunting rifles to deal with. A lot of good men will go down to fight for our country. We wouldn't win, but we'd take down a lot of Americans."

"We did invade you two hundred years ago, in 1812."

"I know, and we won."

TWO DAYS LATER, I DROVE NORTH PAST FORT MCKAY (PRONOUNCED *Mick-eye*) to the edge of the asphalt. There were road signs for Fort Chipewyan and Bear Lake, but that road is only open in winter. There are lakes and rivers that form their own bridges in freezing weather. I stopped along the Athabasca River, downstream from the tailings ponds. The water was muddy along the edge, as would be expected for so large a river. I could not see if it was clear further out. It smelled okay. I placed my hand in the water and wondered what it might be carrying north. I did not know.

On the way back to Fort McMurray, I stopped at a reclamation site near the Syncrude Plant called Gateway Hill. The area was mined and then used as a tailings pond. It was filled with spent sand and planted in 1983 and '84 with green alder, Saskatoon berry, white spruce, aspen poplar and red oster dogwood. The wetlands below the hill were sprouting in cattail, slender wheat grass, red fescue, and gooseberry. The trees on the hillsides were low growing and unspectacular, and the land felt like a dim reflection of its former self, but the breeze was fluttering through the aspen, and birds and insects were flying between branches. There was life here.

The area was quite small next to the vast unreclaimed ponds that surrounded it. But this was a start. The forest recovers so slowly from mining and the tar sands industry is so young that reclamation is very much a learning process in Alberta. Not much of it has happened yet. I spent the late morning hours wandering the two-mile path through the struggling trees, stopping to look and feel what was there. I gave thanks that care had been taken to heal this land. I applauded the people who made the decisions to restore this piece of the earth, and those who carried it out. I do not believe they would—or could—have done this without others shaming them into it. As corporations, they are programmed to extract, to process, and to sell, not to care. Care is a larger organ of ourselves. Care becomes good business only through public opinion. If society did not care, this land would be a toxic mudhole.

But was this land healed? Was this all it takes? After nearly thirty years, what life had returned? Was this a hill before? Were these trees here?

The trees, now looming fifteen or twenty feet overhead, were planted in rows, like cornstalks. How long will this tidiness of aspen and jack pine stand here at attention, stricken by our command, by our sense of order, before natural contours of land enfold them? How long before they look at ease in their native land? Will second or third generations recover the traditional ways? The earth will heal in time, because that is what the earth does. She will return to this place when the tar sands are removed, piped, and burned, but she will not be what she was. She was violated, wounded, traumatized. The scar will show. The wound is deep—through the skin—but the earth is deeper. Her flesh will grow back.

We will be part of the healing process. We will learn from this place. We will watch how the earth heals herself, and we will allow her to heal us. We will find how she moves and move with her. We will come to know her before we ask for her resources, and we will see those who extract and sell from the land as an arm of ourselves and insist they leave an offering. We will respect the earth because she is larger than we. We are in her arms. She is gentle and keeps us in her power.

The woodpecker called faintly from deep within the earth. Dragonflies danced in sunlight along the path.

If the Earth heals without us, our absence will be the healing.

The Carbon Boycott

16

WE ARE AT THE EDGE OF THE FAMILIAR WORLD, STANDING AT A moment of decision ...

We really are standing at the edge of the world of familiar seasons, dependable food supply, and a predictable future. This world will go away soon. Everyone should know this. The Keystone XL pipeline and the Canadian tar sands are not just another environmental disaster, another issue to squeeze into a list of concerns, another thing to think about during the course of a busy day. Now is the beginning of the end of the life we have known.

This could be a good thing. The world we have known—the world of international division and unprincipled economic growth—is no longer viable and will not last whether we are "in favor" of it or not. There is no point in trying to preserve or destroy it. It will pass away of its own accord. Old paradigm questions such as "can we afford an environment?" or "is humanity in our national interest?" will fade away with the worldviews that ask them. But the new world, and the new questions, will come about only with awareness and intention. We will have to make it happen.

As a practical matter, we cannot see and react to fossil fuels as a whole or to climate change in general. They are everywhere, and there is no way to focus on everywhere. Right now, hundreds of supertankers are steaming across oceans, tons of dynamite are blowing tops off mountains, and millions of gallons of toxic chemicals are pouring into hydraulic fracturing operations all over the planet. To concentrate on all of these is to concentrate on none. The irony of climate activism is that we need broad, systemic, global change in how we understand and act, yet we need to focus on specific, tangible issues to create the new consciousness. Finding and developing those issues will test the genius of the new environmental movement.

Climate is not something you can see at any point in space and time. It is an abstraction assembled by thousands of people over dozens of years on every continent of the world. Its substance is tangible, but its expression is mathematical. Climate *change* is an even greater abstraction. It is an assembled picture of many climate pictures. It is an average change in averages, and another purely mathematical expression. Man-made, or *anthropogenic*, climate change is yet another picture that combines the mathematical abstraction of climate change with a derived mathematical figure: cumulative and ongoing fossil fuel consumption by human beings over the last century and a half. The *prevention* of catastrophic climate change—the picture we are ultimately trying to see—opposes the threefold abstraction of anthropogenic climate change with the further abstraction of reducing, eliminating, or reversing human-based carbon emissions in the real world. This is all very heady stuff, and far too big a picture to fit on the average bumper sticker. We need to boil it down.

Here is a boiled-down picture of what the pipeline will do: the Canadian tar sands were recognized in 2003 by the U.S. Energy Information Administration as the second largest oil reserve in world, behind Saudi Arabia but ahead of Iran, Iraq, Kuwait, Venezuela, Nigeria, and the United States. (Venezuela has since been recognized as number one.) If burned, scientists say, the carbon in this tar sand would raise carbon dioxide levels in the Earth's atmosphere to around 450 ppm. The carbon cost of upgrading and refining it bring this figure to over 500 ppm; continued burning of coal, natural gas, and other forms of petroleum at the same time will bring it far higher. This will raise average planetary temperatures by 2 to 3 degrees Celsius[30] within fifty years, enough to exacerbate the ice-melt feedback loop and initiate the carbon cycle feedback loop. (This is in addition to the 1.5 degrees already coming our way from the carbon we've released into the atmosphere already.) Dying plants and trees will release more carbon into the air and remove less from the air as photosynthesis decreases. The atmosphere will then be heating up on its own regardless of what humans do. As the permafrost warms, yet more carbon dioxide and methane will be released into the air. The oceans will warm and rise over coastal areas. Agriculture will become questionable.

This is a middle ground scenario—it may not be so bad; it may be worse. The familiar world outside your window will go into one end of the pipeline and the world of global climate disaster will come out the other.

THE KEYSTONE XL PIPELINE IS THE FUSE OF THE CARBON BOMB.

Thousands of people blocking the right-of-way, delaying construction, even stopping the pipeline altogether will raise the consciousness and solidarity needed for the next step, but none of that will directly reduce carbon emissions. Civil disobedience will not prevent climate disaster in the long run. The way to defeat the fossil fuel industry ultimately will be to dismantle its customer base. People everywhere will have to stop buying what the fossil fuel companies have to sell.

Consumers rule the world. They are the ones who make the final decision. Not politicians, not protesters—*consumers*: people who vote with their dollars to mine tar sands and blow tops off mountains. Consumers are the gas frackers and the pipeline builders. They are the ultimate polluters, and only they can stop pollution completely. We are all they. We are all in this together, and there is no one else to blame.

When you buy an item at Walmart and have it scanned at the checkout counter, a signal goes to a central computer recording the sale and removing the item from inventory. Once the inventory is depleted below a given level, another signal places an order at a factory somewhere on the other side of the planet to make more copies of what you just purchased. The money you spend today determines what gets made tomorrow. The dollar bill slipping out of your hand makes people do what they do all over the world. This happens when you stop at the gas pump or write a check for your utility bill. You may not be thinking about it, but every dollar you spend sends signals into the economy that determine what happens next. The fact that you don't think about it (who does?) means that the economy has no consciousness. Its business is to get people in other places to organize their working lives around getting you what you want.

But it doesn't have to be that way. Consciousness can be injected into the economy by spreading awareness of where dollars go when they are spent. Dollars can be sent deliberately toward good places and away from bad places. But buying consciously is far more difficult than buying what you want when you want. It takes organization, commitment, and discipline. It takes not having what you may want to have and accepting the pain of not having it.

On December 1, 1955, Rosa Parks boarded a city bus in Montgomery, Alabama, and took a seat toward the middle of the bus. As the front of the bus filled with white passengers, she was asked by the driver to give up her seat. She refused and was arrested. This was not important in itself. Other black passengers had been arrested before for the same offense; one of them had boarded a bus at the same stop only nine months earlier. But Rosa Parks's act of civil disobedience became an iconic event of the civil rights movement because it was immediately followed by the Montgomery bus boycott. Community leaders, including E.D. Nixon and Martin Luther King, Jr., asked black citizens to protest the arrest by avoiding the bus system, originally for only one day. (Note their effective use of symbolism: a segregated bus equals a segregated society.)

Their original demands were modest. The existing system required that black passengers fill the bus from the back while whites filled it from the front, but there was no predetermined line dividing the two sections. Whenever a white passenger boarded a fully loaded bus, black passengers in an entire row were required to leave their seats and stand in the aisle further toward the rear. Initially, black community leaders asked only that a line be drawn between white and black sections of the bus so that black people would not have to face the indignity of giving up their seats for white people.

But as the boycott continued and the city felt the financial pain of lost bus fares, black leaders realized the impact they were having and demanded complete desegregation of the bus system. Three-fourths of bus riders in Montgomery were black, and the system depended on their ten-cent fares to remain financially viable. Organizers developed alternative means of transportation for bus riders and kept the boycott alive until they got what they wanted. Former riders rode bicycles, mules, and horse-drawn buggies, or walked or hitchhiked to work. Black cabdrivers agreed to charge the same ten-cent fares the buses charged, and car owners organized carpools. Some whites supported the boycott by driving blacks to and from work. The boycott was mostly a local story but came to national attention when King was arrested and put in jail for "hindering" a bus. Support flowed in from around the United States, and the city was forced to its financial knees. A little more than a year after it began, the boycott ended when the U.S. Supreme Court declared that laws requiring bus segregation were unconstitutional.

The real way to stop the pipeline, tar sands extraction, mountaintop removal, fracking, and climate change as a whole is, then, simply to stop buying coal, petroleum, and natural gas. The biggest businesses in the world will shrivel up and close their doors in a heartbeat, and the problem will be solved for all time. But boycotts work only if alternatives are developed and become available. Organizing other ways to get places was what made the Montgomery boycott work, and only organizing other forms of energy will make the carbon boycott work. So, what other ways are there to do the things that fossil fuels enable—things such as fueling cars, heating homes, and growing food?

Right now, there are few alternatives. To spend no money at all on oil, gas, or coal while remaining part of civil society would be nearly impossible. Everything we do depends on them. Oil is in our coming and going; coal is in our lighting, telephones, computers, refrigerators, and washing machines; natural gas is in the warmth of our houses; and all are in our food. We need petroleum to run tractors and trucks and to make fertilizer; we need coal for grain elevators and electric refrigerators and milking machines; and we need gas to dry wheat, corn, and rice. Each of these fuels is needed separately, and all are needed collectively. Yet we must learn to live without *any* of them.

Is this possible? Of course it is possible. George Washington never used any fossil fuel at all. He never noticed the difference. If he can do it, so can we. Human life is possible without burning any nonrenewable carbon at all. But human *civilization* as we know it is not. We are not what we were in George Washington's day, and we never will be again. Humanity as it is now cannot and will not survive without fossil fuel. We cannot go back. Life does not do "backward" well. We can only go forward. It is comforting to know that if we really wanted, we could live the way people used to live before the Industrial Revolution, but I don't think that will happen, has to happen, or should happen.

Our current use of fossil fuel is not the enemy; it is our vision of *continued* use that is the enemy. The blindness of market forces and the distance of human consciousness from the ecological effects of fossil fuel are the enemy. For the immediate present, we need to keep burning oil, coal, and natural gas, but we need to look at them in an entirely different way. They are *transitional* fuels. They have brought us from the horse-and-buggy days to a higher level of civilization. They have made us who we are, which is, I think, a great deal more than who we were, and we should be thankful for them. At this moment in history, fossil fuels are our ticket to the future—a future without them.

Understanding of fossil fuels will shift as the paradigm shifts. From the standpoint of the ecological paradigm, they become no more than a stage that we are passing through. Like the yolk of an egg or the endosperm of a seed, fossil fuels are stored energy available for civilization to get up and started. Compact and concentrated, they are also finite and exhaustible—perfect natural resources to move humanity from isolated subsistence to a sustainable global society. Coal, gas, and petroleum are easily digestible nourishment for civilization in its early stage. Like mother's milk, they feed the body in its infancy, but they are temporary. The body must be weaned as it matures. The carbon boycott is about weaning human civilization from the mother's milk of fossil fuels, tantrums and all.

Initiating an immediate, cold turkey boycott of fossil fuel would be laughable. It cannot be done without renouncing civilization altogether. But it can be done as a *long-term* movement. There are emerging new ways to heat houses, drive cars, turn on lights, and eat food. Some involve wind, solar, and geothermal energy; some involve *renewable* carbon fuel. (Firewood and biofuels are renewable because they extract as much carbon from the atmosphere in their formation as they release in combustion.) Renewables are pricey and a little geeky at this point, but the more consumers demand them, the better they will get. We are all stuck with the jobs, transportation patterns, food sources, and lifestyles that we find ourselves in, but transitions can be accomplished every time we *change* jobs, houses, cars, or shopping habits. When you look for a new house, buy one or build one close to work and close to public transit. Look for a good, south-facing exposure. Maybe you will have to cut a tree down on the south side, so plant another on the north side. Landscape with solar panels in mind. Ask geothermal and solar installers to check out the site for future potential. Is there a farmers market nearby, or better yet, a good place for a garden? Aim high. Over the years, aim toward coming ever closer to living consciously in the new world. You will eventually become stuck in whatever behavior patterns you initiate, so make them good patterns. Put yourself in a position to honor the boycott as it develops.

Younger people may find it easier to create new lifestyles. Older people may be in a better position to invest in sustainable technologies. Everyone can help boycott carbon!

The key is to stay in the economy while keeping it at a safe distance. You will need money. Money is a good thing—a wonderful thing—that makes it possible for you to do the things you do well without having to do everything. Money secures for you the things other people do well. I recommend it. But keep your spirit out of the economic paradigm by not letting money direct the course of your life. Do with less of it. Don't buy stuff. Fix up old things and learn to do more for yourself. Don't spend all your time making it. Save money to invest in long-term systems that will make you less dependent on everyday money. And be careful of debt; debt creates dependency on the dollar. Try to develop a concept of *enough* money, and live happily within it. The important thing is to put the money economy *within* the larger picture of life.

A garden is the best place I know to get your head out of the economic paradigm. Get out there sometimes, pull a few weeds, haul a little compost, and watch what is happening in your own head. You may find yourself wondering why you are working so hard. Food is still pretty cheap in this country; earning the money to buy a bag of beans takes a lot less work than growing them. That's the dollar talking to you. That's the dollar equation we all carry around in our minds that converts work, food, comfort, happiness, etc., into money. Everything has a price, and the business of modern living is to manipulate money in order to work as little as possible and have as much stuff as possible. That is what makes you a "smart" person. But when you hear the birds sing and you feel the soil between your toes, you will set the hoe aside for a few minutes and make room for higher forms of thought. You will see there are other things going on in the world and in your mind: let them in. That is life calling. At the end of the season you may have a bushel of potatoes, some winter squash, and a few jars of dry beans. You won't embarrass yourself by thinking what they are "worth," but you will know that there is food for the winter. You will live. That much food—that much life—has been demonetized. You boycotted tractor fuel, truck fuel, fertilizers, pesticides, and freezer space at the grocery. You wasted a lot of time that could have gone to making more money than you saved, but you were in direct contact with the earth for that much of the sustenance she provides. She touched you for that much.

The carbon boycott will have three effects. First, it will stop fueling the fuel industry. It will stop pumping money into pipelines, tar sands extraction, drilling, fracking, and mountaintop removal, and it will stop sending signals to fossil fuel companies that we like what they are doing. Second, it will help organize new consciousness. Part of the great success of the Montgomery boycott was the involvement of thousands of common people in a common cause. It brought them together and gave them new vision and new hope. So, do the carbon boycott with your friends and family. Let people know about it. And third, the carbon boycott will ease the transition to a new lifestyle—it will *be* the new lifestyle. Because ongoing consumption of fossil fuel is incompatible with human survival, non-consumption is vital to human survival. In time, *everybody everywhere will boycott fossil fuel.* Learning the best ways to do so will shape the course of civilization. Those who begin the boycott first will show the way.

So let us boycott carbon. But please, don't be a puritan about it. Don't sweat the small stuff; don't do everything perfectly. Don't drive your friends crazy and make them defensive. How your life reflects into the community is more important than any particular action you take. Lighten up. Drive down to McDonald's every once in a while and buy a hamburger. The evolution can wait. Have fun! Enjoy.

August 16, 2012 – Louisville, KY
High 91°F – Low 65°F – Precipitation 0.79 inches

TransCanada has begun construction of the southern leg of the Keystone XL pipeline. Demonstrations broke out in Cushing, Oklahoma, and Dallas, Houston, Paris, Austin, and Nacogdoches, Texas. In an interview with the Los Angeles Times, Ron Seifert said that citizens are prepared to stage sit-ins and other civil disobedience actions to stop it. "Every day we can meet construction with direct action is a small victory for us," he said. "It gives us a chance to describe the harms and abuses and the dangers of this pipeline going forward."

August 23, 2012 – Louisville, KY
High 91°F – Low 62°F – Precipitation 0.0 inches

The Mississippi River is twelve feet below normal in Memphis, Tennessee, thanks to the drought. According to the Coast Guard, nearly a hundred boats and barges were waiting for passage Monday along an eleven-mile stretch of the Mississippi that was closed when a vessel ran aground.

August 28, 2012 – Louisville, KY
High 92°F – Low 70°F – Precipitation 0.0 inches

Seven people were arrested today in Livingston, Texas, protesting construction of the Keystone XL pipeline. I recognize all of them from the training session in Winnsboro. Four of them chained themselves to a truck at the gateway of an equipment yard, blocking the entrance for most of the day. Police had to disassemble the axle of the truck to extract the protesters. The incident was covered on local television.

September 5, 2012 – Louisville, KY
High 92°F – Low 70°F – Precipitation 0.93 inches

Three more activists were detained and later released in Texas today protesting the Keystone XL. TransCanada announced a new route through Nebraska avoiding the Sand Hills. Opponents there stated that the route was only one of their concerns.

September 16, 2012 – Louisville, KY
High 79°F – Low 60°F – Precipitation 0.01 inches

I have decided to go back to Texas. I leave early tomorrow morning to take part in a nonviolent action in the Keystone XL right-of-way near Winnsboro. How can I sit and watch from here?

Epilogue: An Opening Skirmish

WE ARRIVED AT THE SCENE JUST BEFORE SIX A.M. THE AIR WAS cool and clear this September 19, the sky dark and bright with morning stars. A low, grassy hill rose from the highway, strewn with sticks and shredded bark, heavily carved with bulldozer tracks. A dozen pieces of land-clearing equipment were parked at the top—dozers, chippers, a feller-buncher, and log skidders. We walked among them, aiming small flashlights on potential lockdown points and whispering what we knew of their uses. There was time to decide which equipment would be most essential for the day's work, and time to compose ourselves, but there was tension in the air as we thought of workers and police arriving on the site. I walked a short distance from the machinery and stood for a moment looking up at the Orion constellation poised high over the highway; at Sirius, the brightest star in the night sky; and at Venus, brighter yet, hovering nearby. I thought briefly of my new friends along the pipeline in Nebraska, Oklahoma, South Dakota, and far-off Alberta; of the workers at this site in Texas, now rising from their sleep; of the friendly TransCanada vice president I had interviewed over the phone last month; and of the smallness of the earth under my feet. We were just north of Winnsboro.

A lockdown point was found for Gary on the buncher, a huge engine that grabs trees by the trunk and slices them off at the base. Doug and R.C. found a spot at the log skidder and locked together across the track on one side. As the stars faded into the day, we noticed in the dim light that many of the trees in the area were already down. The buncher might not be needed that day. So Gary moved to the chipper. Ramsey was running around snapping photos. All of those avoiding arrest removed themselves from the site. Now, it was time to wait. I stood in the middle of the equipment yard, where the workers would park their trucks. It was my role to talk to workers, and later police, about what was going on—to explain to them that the protesters were safe where they were, that we were doing no damage, and that we were not protesting *them*. We were protesting the pipeline they were building. I was to keep the temperature down and do all I could to keep things safe. I was risking arrest in the performance of my role, but I planned to leave the area when warned. We waited. I walked back and forth between the chipper and the skidder checking on the three lockdowns, making small talk to ease the tension.

Pickup trucks began arriving on the far side of the road as daylight filtered through the stars. I stood on the hill where people could see me. A TransCanada truck drove several yards past me and stopped. Nobody got out. I moved up to between the truck and where Gary was locked down, and we talked quietly for a few minutes. A man in a white TransCanada hardhat emerged from the truck and I greeted him. The back of his jacket read *Dig Inspector*.

He returned the greeting and said, "Unless you have permission from the landowner, I am asking you to leave."

I said, "Thank you. I know," and did not move.

"Nothing here is worth anyone getting hurt," he added.

"I agree," I stated. "We are sorry to inconvenience you. We are not protesting against you personally. But we are opposed to the project you are working on here."

"I have called the sheriff," he informed me. "They are on their way. Please be careful."

"Thank you." I paused. "It is a beautiful morning."

"Yes, it is." He smiled and returned to his truck.

More men in pickup trucks were showing up on the other side of the highway but keeping their distance. As sunlight poured in over the eastern tree line, the sheriff arrived. I stood still as he walked up to me. There was attitude in his pace.

"This is your warning, now *leave*!" he said, pointing down the hill.

"This protest is not against you or the workers here," I replied.

"Do you ever think about the emergency calls we should be making?" he said.

"I am sorry to be inconveniencing your day. I wish there were another way to do this." He ignored me and stormed off to where Gary was chained to the chipper. For us, this was an opening skirmish in the long struggle for habitability of the home planet; for him, it was mischief.

I could stay where I was in the yard and get arrested, but I did not think that would be of great importance. So I went over to Gary and then to Doug and R.C. to see if they were all right and to explain that I had to leave. They were fine, so I walked back down the hill to where Ramsey was standing near the highway. The sheriff indicated that that was not far enough away, so we crossed to the far side of the highway, where we could still see the other three, however distantly. Other police, including two female officers, were walking around the yard, trying to figure out what to do. A man came with a hack-saw. The sheriff walked past us several times. It was unclear what he meant by 'leave,' but I needed to keep watch over the three lockdowns and felt that his lack of reaction to us was tacit approval of our standing off the Keystone property, several feet from the pavement, but well within the public right-of-way. Traffic moved freely along the highway. Ramsey was taking pictures and texting on his cell phone.

We stood there for close to an hour, watching the police extractors trying to break through the lockboxes. A cloth sheet was placed to cover the area where they were working on Gary, and a pickup truck was parked to block our view of Doug and R.C. I was avoiding eye contact with the sheriff.

Then, from nowhere, he appeared in front of us and said, "I told you to leave. You're under arrest." He cuffed Ramsey and a deputy came at me from behind and cuffed my wrists. "It was unclear to me where you wanted us to go," I said to the sheriff.

"That's pretty obvious," he muttered, and they led us away.

THE HOLDING CELL WAS A LONG CONCRETE ROOM WITH METAL benches along each wall and a toilet at the far end. There were no cushions, windows, televisions, clocks, or reading material. All personal possessions were taken from us. Ramsey and I were ushered into the room around 9:30 in the morning. We sat there in prison stripes for an hour or two before the door opened and Gary and Doug walked in. R.C. was kept in a women's cell. All three of them had been extracted from the construction equipment one way or another, but they were safe and unhurt, except for a bruise on Gary's forearm. Around midday, white bread, peanut butter, and sweet tea were served. We sat for hour after hour through the afternoon. In the early evening we were called, one by one, to the front desk. The clock in the hall read 7:30 when my name was called. Normally, inmates are told of the charges against them within a few hours and then released with bail or under their own recognizance. We would be held overnight.

Doug, Ramsey, and Gary were in the overnight cell when I arrived with my mattress and blanket. We greeted, cheered, and high-fived, congratulating each other for graduating from the holding cell. Here, at least, we would have bunks and, for better or worse, a television. We turned on the television and watched ourselves on a local news broadcast.

We were in the cell that night, the following day, and the following night. Nothing was given to us to do. But it turned out that I have good jail skills: meditation, yoga, running in place, and pacing, interspersed with meals, naps, news programs, and chats with my cellmates. It wasn't so bad. We did not know we would be there the first night, or the second night, or that we would be held until late morning the third day, but I tried not to worry about being released. Occasionally, at the bottom of my mind, I heard a queasy, rumbling voice say things such as "When will we get out?" and "I'm going nuts in here." I listened but did not respond. I knew we would get out, that there were people watching on the outside, and that ours is a society that believes in the rule of law. I was glad to be in America and not some forgotten dungeon, or some backward society that dealt with troublemakers like me in summary fashion. I was glad not to be in Guantanamo Bay or Abu Ghraib, where people who believe in human rights do not extend them to non-American humans.

The cell was clean, fairly spacious, and bordering on comfortable. Once we got past the peanut butter, the food was pretty good, too. The jail personnel were helpful, polite, and humorous. They became quite relaxed around us—I think we provided them comic relief from their regular clientele. An attendant was in the cell talking to us at one point when I noticed that he had left the door open. They brought us tea in the middle of the afternoon one time, just to be nice. The taxpayers of Franklin County would not have approved of any more amenable accommodations; we might have booked another night. But the shower was cold, there was no soap at first, and we had no way to brush our teeth. We walked out in the same socks and underwear we walked in with.

Ramsey found a broken pen and some scrap paper that we all used to write essays, poems, and letters to the editor. We noticed that there was a checkerboard lightly scratched into the metal table in the center of the cell. It had been painted over several times and was difficult to see, but it remained visible where light from the ceiling glanced off the table at a wide angle. We made up paper knights, pawns, kings, and queens, and we began a chess tournament that lasted throughout the stay.

Early Friday morning, we were paraded in front of a magistrate at the jail and charged with "obstructing a highway or passageway," a Class B misdemeanor. We were held without charges for just under the forty-eight hour limit, but then we were returned to our cell for several hours more, without any idea what was happening on the outside. We were released late that morning.

A half dozen or so Tar Sands Blockaders were waiting for us outside. We all whooped and hollered, took pictures, and held banners in front of the jailhouse. It was a gorgeous fall morning, and I was elated to be free in the sunshine.

Franklin County was through with us, but TransCanada was not. We were each handed a citation from the sheriff's office with our name listed as defendants in a civil suit. There were fifty-two pages of descriptions, numbered paragraphs, photographs, exhibits, declarations, and a chart of the equipment yard showing the Keystone right-of-way. Ramsey and I are not listed as attached to machinery, but the document claims that we refused to leave the property, which is not true. They do have a photo from the website of me holding a banner in front of the machinery, which confirms my presence on the property before I was asked to leave. Somebody worked all day and all night and all the next day on this citation, which is why, I realized, we were held for so long. I am reasonably convinced that TransCanada prevailed on Franklin County to keep us locked up while the citation was being prepared. I am aware the case is serious, but I am not especially concerned about it. No matter the legal entanglement, my conscience is clear. I would do this again.

WHERE ONLY RON SEIFERT AND I HAD STAYED IN THE CAMPING area on David's land back in July, there were now a dozen or so tents scattered through the trees and around thirty people in residence. I walked down the path to the creek and through the tree village to the north border next to Becky and Sam's land. A row of forty-foot tall poles were sunk in the ground, about eight feet apart, spanning the width of the right-of-way. Mike was on a ladder about thirty feet up, securing wooden planks to a catwalk along the wall. Buffalo, who had organized our action, was pointing to a plywood platform swinging directly overhead, explaining how, at a critical moment, he could move from the catwalk to the pod. Ditches were dug on the north side of the wall and heavy timbers buried up against it. The structure had the feel of a primitive frontier fortress, a last-ditch battlement to stem an oncoming barbarian invasion. But there would be no bullets or arrows, no slings or catapults, only young lives clinging to the catwalk, watching the machinery roar below.

As evening fell, a council was called around the campfire a few feet away from the tree village. The subject was the roles that various people would play in the coming siege. Four groups of blockaders were recognized: sitters (trees and wall), rovers (mobile reconnaissance and resupply), "Hot Camp" (the existing encampment), and "Cold Camp" (a new, more secure encampment at a safe distance). The existing encampment was no longer secure, even though it was clearly off the Keystone right-of-way and clearly on David's private property. Surveyors had already come and seen the tree shelters, a TransCanada helicopter had been hovering over the tree village, and the police knew something was up. They were already arresting people such as Ramsey and me for being *near* the right-of-way. It was decided, therefore, to move organizers, training, supplies, and other supporting functions to Cold Camp. Hot Camp would continue but primarily for viewers and media. It would no longer be used as a base camp. A "firewall" would be established along the border between the camp and the right-of-way, cutting off direct contact with the tree village. Convenience would be sacrificed for security. This would reduce liability for David and his family and for witnesses visiting the action.

The supply tree that was connected by trolley to the other trees in the village could no longer be depended on for lifting food and water to the sitters. This is the tree I climbed with David back in July. It was well off the right-of-way, but police were likely to keep people away from it during the siege. Black-painted buckets might still be used at night, but rovers would be prepared to enter the tree village from the west side (opposite Hot Camp) and quietly resupply each tree shelter from the ground. Rovers would avoid direct contact with Hot Camp, camping out elsewhere in the woods, while keeping track of construction progress along other points on the route. They would also assist in bringing other blockaders through the woods from Cold Camp for ground actions below the tree village. The siege is likely to begin in the next few days.

September 22, 2012 – Winnsboro, TX
High 91°F – Low 60°F – Precipitation 0.0 inches

I slept in my car near the tree village last night and left in the morning dark. As I turned onto the highway a half-mile north of David's house, I saw a TransCanada truck parked where the pipeline right-of-way crosses the road. There, in the early light, three men stood looking across the highway, one of them pointing south through the woods toward the tree village.

I BELIEVE THIS IS AN OPENING SKIRMISH IN A LONG STRUGGLE—a war, perhaps—between those who believe that the environmental crisis is real and those who promote economic growth at any cost. The Keystone XL pipeline is the focal point of a larger conflict of worldviews. The upcoming battles will be fought not with fact and reason versus ignorance, but with grit and determination versus financial power. I do not believe a negotiated settlement is possible.

Trespassing, obstruction, jail, and cat-and-mouse with local authority are distasteful to me personally. I do not enjoy this line of work. I prefer the use of research, logic, and reasoned argument in approaching social and ecological issues. I wish that politicians and business leaders were open to the evidence that continued fossil fuel combustion is incompatible with human life on Earth, and I wish that I were still open to the notion that material production is the primary goal of human society. But they are not and I am not. Each worldview is built of a separate rational structure, and there is very little ground between them.

If this is war, it is a peaceful war. There will be no resorting to violence or destruction of any kind. There will be no enemies. There will be no dehumanization of the opposition. Civil disobedience will remain civil at all times: our bullets and bombs will be kindness and compassion. Our anger will be polite and contained. There will be no cities captured, no territories taken; hearts and minds will be the only prize.

It will be hearts and minds that judge us, and themselves; it will be they who serve the sentence or walk in the light.

September 24, 2012 – Winnsboro, TX
High 91°F – Low 66°F – Precipitation 0.0 inches

Land-clearing equipment has moved in from the north where the right-of-way crosses the highway. Trees are falling in Becky and Sam's land next to David's woods, and eight sitters have climbed up into the trees. The battle has begun.

Notes

1. Nikiforuk, Andrew, *Tar Sands: Dirty Oil and the Future of a Continent*, Greystone Books, Toronto, 2010, pp. 15–27. Many of my figures on the mining and upgrading of Canadian tar sands come from this book; it is a standard work on the subject.
2. Nikiforuk. Pp. 132–133. These figures assume that the Keystone XL is built, and that oil derived from tar sands reaches the world market.
3. Lynas, Mark, *Six Degrees: Our Future on a Hotter Planet*, National Geographic, Washington, D.C., 2008, p. 279 (chart). This book describes the consequences of climate change from 1 through 6 degrees Celsius of warming, including which positive feedback loops are likely to begin at which temperature levels.
4. Kuhn, Thomas, S., *The Structure of Scientific Revolutions*, Second Edition, Enlarged, The University of Chicago Press, Chicago, 1970 (1962). This is one of the more influential books of our time. Kuhn's concept of the paradigm has been so successful in revolutionizing understanding of scientific progress that its use has spread naturally into other areas, including this book.
5. Darwin, Charles, *On the Origin of Species*, (authorized edition from 6th English ed. New York, 1889), II, pp. 295–296. Quoted from Kuhn, p. 151.
6. Planck, Max, *Scientific Autobiography and Other Papers,* trans. F. Gaynor, New York, 1949, pp. 33–34. Quoted from Kuhn, p. 151. Both Planck and Darwin knew they were speaking more to coming generations than to their own colleagues.
7. "TransCanada to push ahead with part of Keystone pipeline," *Washington Post*, February 27, 2012.
8. Studies by the Natural Resources Defense Council and the Cornell University Global Labor Institute show that the Keystone XL will actually *increase* gas prices in the Midwest by diverting Canadian crude oil from refineries there to the Gulf of Mexico.
9. Kintisch, Eli, *Hack the Planet*, John Wiley and Sons, Hoboken NJ, 2010. A forest seems to be more than the sum of its trees.
10. Lynas, p. 138.

11 Lynas, p. 279. This refers to the same chart as note 3. Positive feedback loops cannot be fully isolated from one another. As one of them raises global temperatures, even slightly, it is likely to initiate another. "Runaway" climate change happens when one feedback loop raises temperatures past a threshhold point that starts the next one. Many scientists believe we are close to a tipping point that may begin a series of such feedback loops. If the Alberta tar sands are mined and burned, we will almost certainly find ourselves in the early stages of a runaway climate.

12 Lynas, p. 178, from a quote from John Sheehy of the International Rice Research Institute in Manila.

13 Lynas, pp. 47–94.

14 Hertsgaard, Mark, *Hot: Living Through the Next Fifty Years on Earth*, Houghton Mifflin Harcourt, 2011, p. 250. The German Advisory Council brings the global picture to the household level. If we are interested in a good chance of staying within two degrees of warming, it is plain that we are each spending much more carbon than the budget calls for.

15 Wallace, Scott, *The Unconquered: In Search of the Amazon's Last Uncontacted Tribes*, Crown Publishers, New York, pp. 426–427. (Wallace, "Last of the Amazon," *National Geographic* 211, no.1, January 2007, p. 49. And: Simon L. Lewis, et al, "The 2010 Amazon Drought," *Science* 331, February 4, 2011, p. 554.)

16 Hertsgaard, p. 13.

17 Common Dreams, April 12, 2012.

18 I spell out this perspective more thoroughly in Avery, Samuel, *The Globalist Papers*, Compari, 2005.

19 Committee on Energy and Commerce, U.S. House of Representatives. April 18, 2011.

20 Colborn, Theo; Kwiatkowski, Carol; Schultz, Kim; Bachran, Mary (2011). "Natural Gas Operations from a Public Health Perspective" (PDF). "Human and Ecological Risk Assessment: An International Journal" (Taylor & Francis) 17 (5): 1039–1056. DOI:10.1080/10807039.2011.605662. How or why fracking ever became *exempt* from the U.S. Safe Drinking Water Act is entirely beyond my comprehension.

21 "Local zoning provisions in Pa.'s gas drilling law." *USA Today*. Associated Press. March 3, 2012. Retrieved February 23, 2012. These provisions were later rescinded, as I report later in the narrative.

22 Sheppard, Kate. "For Pennsylvania's Doctors, a Gag Order on Fracking Chemicals. A new provision could forbid the state's doctors from sharing information with patients exposed to toxic fracking solutions." *Mother Jones*. Retrieved March 23, 2012. Getting this through the state legislature must have required some very creative lawyering.

23 Maugeri, Leonardo, *The Age of Oil: The Mythology, History, and Future of the World's Most Controversial Resource*, Westport, Connecticut, Praeger Publishers, 2006, pp. 212–216.

24 "USDA Declares 'Natural Disaster' in 26 States as Drought Devastates," Common Dreams, July 13, 2012.

25 *Rolling Stone,* Politics, July 19, 2012.

26 Weaver, Andrew. "New Study: Coal is 1500 Times Worse for the Environment than Oil Sands," Huff Post Science, posted February 21, 2012. (First appeared in *Nature Climate Science.*) Wishart quotes the study accurately. The point of the article is to suggest that coal is much worse than tar sands, though the study does not include greenhouse gases emitted in the extraction, refining, and transportation of tar sands. I have no idea how Weaver came up with these numbers (the article does not say), but they are much, much lower than anything I have heard from any other source. What is interesting to me here is that Wishart has chosen a set of figures that fit into a constellation that supports TransCanada'a case, where I have chosen others to support mine. I have every reason to believe that mine are closer to "objective fact," but they nonetheless reflect my overall worldview.

Despite TransCanada's use of his figures, Weaver by no means supports tar sands development or the Keystone XL pipeline. "I have always said that the tar sands are a symptom of a bigger problem. The bigger problem is our societal dependence on fossil fuels." He concludes that, "... as a society, we will live or die by our future consumption of coal. The idea that we're going to somehow run out of coal, natural gas, and other fossil fuels is misplaced. We'll run out of our ability to live on the planet long before we run out of them."

27 "DeSmogBlog," July 27, 2012.

28 "Nation of Change," August 6, 2012.

29 Schulte, Grant, "Thousands of Fish Die as Midwest Streams Heat Up," Associated Press, August 2012.

30 Lynas, p. 272. "Climate sensitivity" is a term scientists use to relate increased carbon levels in the atmosphere to increases in global temperatures. Computer models indicate that a doubling of carbon levels from 280 parts per million (the preindustrial level) will increase temperatures by about 3 degrees Celsius. This is the amount of carbon that will be released in the development of the Canadian tar sands; the expected temperature increase is the likely threshold point for the carbon cycle feedback loop, and possibly the permafrost feedback loop as well.

Index

"This logo identifies paper that meets the standards of the Forest Stewardship Council®. FSC® is widely regarded as the best practice in forest management, ensuring the highest protections for forests and indigenous peoples."

Ruka Press is committed to preserving ancient forests and natural resources. We elected to print this title on 100% post consumer recycled paper, processed chlorine free. As a result, for this printing, we have saved:

31 Trees (40' tall and 6-8" diameter)
14 Million BTUs of Total Energy
2,696 Pounds of Greenhouse Gases
14,623 Gallons of Wastewater
979 Pounds of Solid Waste

Ruka Press made this paper choice because our printer, Thomson-Shore, Inc., is a member of Green Press Initiative, a nonprofit program dedicated to supporting authors, publishers, and suppliers in their efforts to reduce their use of fiber obtained from endangered forests.

For more information, visit www.greenpressinitiative.org

Environmental impact estimates were made using the Environmental Defense Paper Calculator. For more information visit: www.papercalculator.org.